KEEPING BEES
IN TOWNS & CITIES

"Instead of dirt and poison, we have chosen to fill our hives with honey and wax, thus furnishing mankind with the two noblest of things, which are sweetness and light."

– Jonathan Swift

KEEPING BEES

IN TOWNS & CITIES

LUKE DIXON

TIMBER PRESS
London · Portland

Title page photograph
Hackney City Farm, London

Copyright © 2012 by Luke Dixon. All rights reserved.

Published in 2012 by Timber Press, Inc.

2 The Quadrant	The Haseltine Building
135 Salusbury Road	133 S.W. Second Avenue, Suite 450
London NW6 6RJ	Portland, Oregon 97204-3527
timberpress.co.uk	timberpress.com

Printed in China
Second printing 2013

Library of Congress Cataloging-in-Publication Data

Dixon, Luke.
   Keeping bees in towns and cities / Luke Dixon. —1st ed.
       p. cm.
   Includes index.
   ISBN 978-1-60469-287-7; 978-1-60469-378-2
 1. Bee culture. 2. Urban agriculture. I. Title.
   SF523.D58 2012

   638'.1—dc23

               2011050645

A catalogue record for this book is also available from
the British Library.

CONTENTS

Introduction   7

Boxes of Bees   19
SELECTING THE PERFECT HIVE

Buying Bees   35
THE GOOD, THE BAD, AND THE SWARMY

Life Inside the Hive   53
LOOKING AFTER YOUR COLONY OF URBAN BEES

Sex and the City Bee   69
THE LIFE CYCLE OF A HEALTHY COLONY

The Nectar Garden   85
FROM PLANTS TO HONEY

Honey Harvest   99

Winter   113

Reports from the Field   121

Resources   176

Index   178

Acknowledgements   182

"If you have room for a composter or a water barrel, you have room for a beehive."

INTRODUCTION

A life lived in the a city can be one far away from the concerns of the natural world. The day is dictated by the clock and the commuter timetable, the light is artificial, we are sheltered from the weather, and underfoot concrete and tarmac are hard against our feet. It is easy to imagine that the wild world has been kept at bay, far from the urban sprawl. But the untamed creatures are there as they have always been, and as our towns and cities have grown and man has become an urban animal, so those other creatures who have always lived on the land have found different ways and places to live as man has encroached—foxes, badgers, mice, and others much less obvious have all adapted to city life. So have bees and there can be few better ways of maintaining a connection with the natural world than by keeping a hive in the middle of the city.

Man and honey bee have always lived in close proximity. As man has felled the trees and built over the land, the bees have had to find other places to make their homes and have moved into chimneys and wall cavities, church spires and attics. As long as there have been towns there

The Paris Opera house

have been urban bees, and as long as man has been keeping bees there have been urban beekeepers.

Until recently there has been something secretive about the keeping of bees in towns and cities. Anxious to avoid fear and censure, people have kept their bees for the most part away from public gaze, very often up above the city streets on roofs and balconies. Some countries have laws against it (no one can keep bees inside the city limits of a Spanish town). In New York the keeping of bees was forbidden until the spring of 2010 when public anxiety at the plight of the honey bee saw the anti-bee statutes

revoked, though guerrilla beekeepers had been quietly tending hives in the city for decades before.

Londoners have always kept bees. There has been an association for beekeepers in the city since 1883. A century ago there were an estimated one million managed colonies of bees in Britain. Now the figure is a quarter of that but is growing again. It has been the decline of the honey bee and our increasing awareness of our dependency on it that has been in large part responsible for the growing twenty-first-century interest in bees and beekeeping. As England became urbanised, bees

Far left
*The roof of the barracks
of the Port-Royal in Paris*

Left
A balcony in Paris

Right
A Paris garden

became fewer. Before the industrial revolution some think that every second or third dwelling had a beehive. As the cities encroached on the countryside many of those hives disappeared. Others just moved upwards onto roofs.

Renewed interest in urban beekeeping had yet to take root when I began keeping bees a decade ago. I had spent my life till then working in the theatre, indoors in the darkness, making plays and sometimes going for weeks on end without seeing daylight. It was the need for something quieter, more peaceful, more 'natural' in my life that reawakened my interest in bees, a fascination I can recall when as a small child I watched the bees in an observation hive in our local museum in north London. The image of the bees flying in and out through the small hole in the glass window which gave them access to the hive from the outside world has stayed with me ever since. I remembered those bees when many years later I was searching for a hobby that would take me away from the often irksome human egos of the rehearsal room, to the open air and something more contemplative.

I searched around and found an evening class in beekeeping. Once I had finished the short course, met many of London's urban beekeepers, and become confirmed in my desire to have a hive or two of my own, I had to find somewhere that I could keep them.

New York rooftop apiary

Left
Beehives at night in Coram's Fields, London

Right
WBC hive close by The Monument in the City of London

Living in the middle of Soho, in the very heart of the West End of London, I found my neighbours none too enthusiastic about my plan to put a hive on the roof. The members of my local beekeepers' association, who had taught my course, had their hives in all manner of places. Most were in gardens large and small, many were on allotments, and some were on rooftops. It seemed that you could keep a hive pretty well anywhere. As I met more London beekeepers I found hives in ever smaller spaces—wedged onto balconies on council flats, outside French windows in minuscule Victorian gardens, and perched on roofs. It was clear that if you

have room for a composter or a water barrel, you have room for a beehive.

As I visited the hives of others, I learnt what to consider in choosing a suitable site. You do not need much space, just enough for the footprint of the hive and room around it for the beekeeper to work without being cramped and with space enough for the bits of the hive when it is taken apart to inspect the bees within. You need some security. The biggest threat to the urban bee, I was told, was the urban vandal. Otherwise the bees can look after themselves. Like us, they need a home and plenty to eat and drink. If the beekeeper provides the home, the bees will seek out pollen and nectar to feed

on and find sources of water to drink. The city offers more variety of forage than the countryside ever can, which, combined with its warmer climate, means that the urban environment is perhaps better suited to the modern bee than the rural environment is. The more beekeepers I met, the more I realised how straightforward urban beekeeping could be if only you have a place for your hives.

I began my search and the further I looked the more potential hive sites I found. The inner city is full of wild sites, of green and brown land neglected and unused. Wastelands around railways, motorway embankments, strips of meadowland by reservoirs and sewage works, patches of green on industrial estates, and parks and gardens. So many parks, so many gardens—all rich with forage and plants that need bees to pollinate them. I was told that a good rule of thumb is two hives for every acre of forage. As I explored London on my scooter, I realised that there were thousands of acres of forage just waiting for bees.

It took a while to persuade someone

to let me put my hives on their little bit of urban greenery, and I spent many weeks in search of a friendly site owner. It was Caroline Ware, manager of the Wildlife Garden at the Natural History Museum in South Kensington, who offered me my first site. She had been looking for a beekeeper for the garden and offered to let me keep my hives there. The site had everything a bee and beekeeper could want. There was space for two hives just far enough away from public paths to be secure. The garden, though small, was planted so that each area represented a different natural habitat: there was meadow, hedgerow, chalk downland, heath and woodland, even a tiny fen. There was also a pond, so water was not going to be a problem. And if the bees got bored with the forage near to home, there was the vast expanse of Hyde Park a short beeline away.

My first season was idyllic. As I hammered bits of wood together to make my initial hives, I learnt new skills I never thought myself capable of. Once the first hive was finished, John Chapple, an older and wiser beekeeper, brought along a

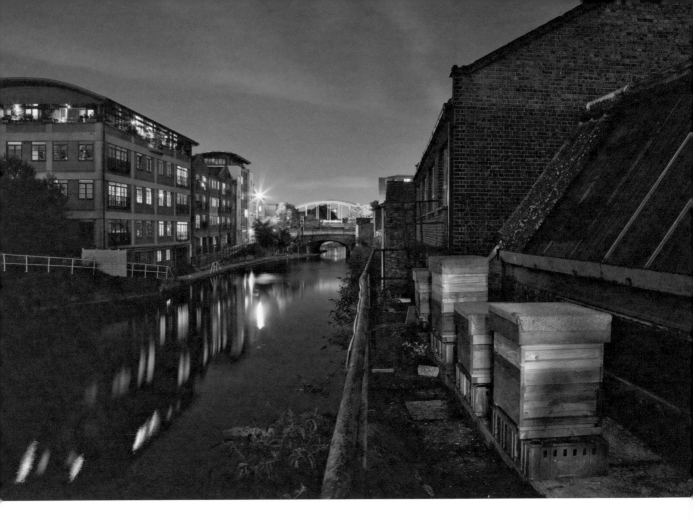

colony in a box amongst the beekeeping clutter in the back of his car, and we transferred the thousands of grumpy and confused bees into their new home. In the coming months John became my mentor, and his experience and knowledge made my early attempts at keeping bees much more confident than they would ever have been had I been on my own.

After we had put the bees into the hive, I sat in front of the hive entrance and waited. Within moments the first bees began to emerge into the sunlight, flying around to literally get their bearings. It was not long before bees were off flying from the hive and coming back laden with brightly coloured pollen. That winter I had a kitchen full of jars of honey and the Wildlife Garden had more berries on its holly trees than ever before. It was a memorable Christmas.

Today the plight of the honey bee is well known and we are all more aware of how dependent we humans are on them for pollinating much of our food. Support for urban beekeepers could not be stronger. About a third of all the food we

Regent's Canal, Hackney, London

eat and drink is the result of pollination by honey bees and without them our diets would be much the poorer. Einstein is reputed to have said that if the bee disappeared off the surface of the globe then man would have only four years of life left. And it would be a life spent eating porridge.

I still work in the theatre. It is the perfect winter job, indoors and warm while the bees are in the hive clustered around the queen and waiting for spring. I have become, like them, attuned to the weather and the comings and goings of the seasons. The bees seem to intuit changes in the weather, while I have to keep an eye on the thermometer and have the Weather Channel in the corner of my computer screen. As soon as it is fine enough for the bees to be out of their hives, I too am out and about at my urban apiaries, clambering over roofs and looking out over all the green spaces spread out before me—parks, gardens, allotments, window boxes and hanging baskets, green and brown roofs. A feast of bee forage in the urban jungle.

TOP TIPS

GETTING STARTED IN BEEKEEPING

- You need a footprint of about 60 cm square for a box frame hive. A long top bar hive with its splayed supporting legs, needs a footprint of about 100 cm square. And you will need some space to work behind and beside it.

- Make sure you have got easy access for getting hive parts and bees to and from the hive.

- Think about your neighbours.

- Cats and dogs seem to co-exist quite happily with bees. Bees and horses do not make good neighbours.

- Face the hive away from humans (pathways, windows, children playing) so the bees can fly out and off from their home without disturbance.

- Provide water for your bees if none is nearby.

- Find a mentor or a bee buddy to help you.

- Have two hives if you can; you will be glad of the spare as your beekeeping and your bees develop.

"All these many types of hive share one thing in common: they are just boxes for bees."

BOXES OF BEES

SELECTING THE PERFECT HIVE

Having found my site, I had to decide which hives to buy—the choice was bewildering. Man has been keeping bees for thousands of years and through all that time has been in search of the perfect beehive. And through all that time, bees have remained wild creatures, 'kept' only as long as they have chosen to be. In the wild, bees live in dark voids, usually empty spaces inside dead trees, sometimes in caves or mountain clefts. They are creatures of the dark, the queen in the deepest darkest space of all, the tens of thousands of workers around her communicating not by sight but by smell, a complexity of pheromones that tell the colony all they need to know and control all that they do. Bees have been on our planet for something like 130 million years, as long as the first flowering plants. Flowering plants are dependent on flying insects for pollination and flying insects are dependent on the nectar and pollen in flowers for food, so the two must have evolved side by side. It was a long while before man came along, and for most of those 130 million years bees lived without humans stealing their honey. The first

Johannes Paul, inventor of
the 'Beehaus', on his roof
in central London

Left
The Spanish cave painting

Above
Skeps in a bole

hominoids—the 'homo' superfamily—
came along about two and a half million
years ago. Humans as we know them,
the species *Homo sapiens*, appeared about
160,000 years ago. For thousands of years
humans did not think to 'keep' bees but
left them in the wild, raiding nests for
honey and wax.

On the wall of a cave in southern
Spain is a painting. It shows a figure
climbing to a bees' nest and stealing
honey. It was painted perhaps three
thousand years ago and we will never
know why or by whom. It is the earliest
image we have of human and honey bee.
Sometime between the appearance of
man and that image being painted onto
the wall of the Spanish cave, humans
realised the taste and energy-giving
properties of honey. They may also have
discovered the wonderful light that could
be made from the burning of beeswax.
Our Spaniard is not keeping bees but
climbing high into a tree to retrieve

Straw hive in Gambia

honey and wax from a wild nest. In some parts of the world that is still the way in which honey is taken, and bees are not kept but left to live wild as they have always done.

The first beekeepers

It was all very well climbing high into trees and up cliffs and crevices, but what if the bees could be kept somewhere closer to home and contained in something from which the honey and wax could easily be extracted? Eventually humans devised ways to keep bees, breed them, and harvest wax and honey as it was needed. The first hives replicated the bees' homes in the wild, carved out of felled tree trunks and placed horizontally or vertically with lids or doors for easy access to the colonies within.

There are other early images of man and bee and archaeological discoveries of ancient beehives. Pictures of hives made of clay can be found in tombs of the pharaohs; such hives are still used in the Middle East. Other cultures wove their hives from straw or wicker. You can still find these skeps, as they are called, in some parts of Europe. They are like upside-down wastepaper baskets with a small hole for the bees to come in and out of, and they have to be brought indoors or stored in a brick 'bole', a kind of open-air cupboard, over cold winters. Different designs of hive were made in different parts of the world, using the materials to hand and adapted to different climates. All these simple hives have a major disadvantage—the colonies of bees have to be destroyed in whole or part to extract their contents.

The birth of the modern beehive

It was an American pastor, the Rev. Lorenzo Langstroth, in Massachusetts in 1851, who came up with a solution. Bees, he noticed, always build their comb in vertical sheets and always with just enough space between them for the bees to pass one another on facing sheets of comb without knocking each other off. This is the crucial 'bee space'. Too much space between the combs and the bees

BOXES FOR BEES

Langstroth

National

WBC

will fill it with wax, too little and the bees will not be able to work. If the parts of a hive could be kept between ¼ and ⅜ of an inch (6 to 9 mm) apart, the bees would be able to work and would not join them together with bits of comb or sticky propolis, a kind of super-glue made from tree and plant sap. Langstroth realised that if the bees could be persuaded to build their comb on moveable wooden frames that could be kept in boxes piled on each other and removed at will without destroying the nest, it would then be possible to inspect the bees, remove frames and extract honey, add

extra boxes of frames to encourage honey production, and so on—all without harming the colony. The story goes that Langstroth experimented with an old Champagne crate and that it was the size of the crate that dictated the size of his frames. Whether true or not, Langstroth created the first modern beehive, and all those made since have been constructed on his principles.

There is a myriad of variations. In the United States, hives are still built to Langstroth's specifications. Langstroth's hive was first mass-produced by the A. I. Root Company. Amos Ives Root offered

Beehaus   *Warré*   *Top bar*

them free to Christian missionaries and they spread around the world like Gideon Bibles. Other countries adopted various modifications in shape and size.

The Langstroth hive itself, big, strong, and sturdy, remains the U.S. standard. The even bigger Dadant is also a popular American hive. English beekeepers mostly use a National hive, smaller and more compact, or the WBC hive of William Broughton Carr, smaller still but encased in an elegant shell of slanted white-painted slats that keep the hive warm in winter, help reduce the temperature on a baking hot summer's day, and protect the boxes inside against the elements. In Scotland, where the hives are moved around to follow the flowering of the heather and the much-sought-after heather-flavoured honey that comes with it, the Smith, smallest hive of all, is the most popular.

All these hives involve piling boxes one on top of the other. Japanese hives, and 'The People's Hive' invented by another pastor, the Frenchman Abbé Émile Warré, have smaller boxes which are added underneath one another, allowing the bees to extend their comb ever downwards.

Langstroth's frames have also been adapted to go in top bar hives that are very like the old horizontal tree trunk hives, the frames hanging in a long row and the bees building outwards, one comb next to the other, rather than upwards. For a century and a half, all hives were made from timber. By the time I had to choose which hives to buy, I not only had choices of wood but hives available in plastic and polystyrene. All these many types of hive share one thing in common: they are just boxes for bees.

Early beekeepers kept their bees for honey. Later the wax became as, if not more, important, beeswax candles providing the only clear, smoke-free source of light for churches and cathedrals. As human agriculture developed, so too did husbandry of bees for pollination. Today we keep bees for those practical reasons, but also more altruistically to do a little something towards the health of the planet, or that little bit of the planet on which we live. And you can keep bees for no other reason than to stay in touch with the natural world in the middle of the city. There are hives for every type of beekeeping—the modern hive, a pile of boxes for honey production; long hives and 'natural' hives where the health of the bee is more important than harvesting honey for humans; ornamental hives and utilitarian hives; hives of wood and plastic and polystyrene.

Choosing a hive

My first task having been offered a home at the Natural History Museum was to decide which of these many types of hive to use, and just as importantly what my hives should be made of. My site was in the very middle of London, at one of the city's most popular tourist destinations. As an inner city urban beekeeper I had only a very small space available for my hives, which ruled out the long top bar hives, and also the very English WBC, which needs a lot of space around it when it is being taken apart. Like many other novice beekeepers my choice was also greatly influenced by my fellow local beekeepers. I was lucky in having a hugely experienced mentor in John

Bee on author's weathered cedar hive

Chapple. He used National hives and they were by far the most common amongst the London beekeepers from whom I might have to beg and borrow parts, and whose hives I would be able to look into to learn more about my new hobby. National hives it would be.

I was also keen on having wooden hives. The hives would be on public display, with the public passing very close by them. So it was in part an aesthetic choice, but there was also something about the look of a wooden hive which made it immediately recognisable as what it was—a box full of bees. For me the new plastic and polystyrene hives did not advertise themselves as beehives, and it was important that my hives were obvious to anyone coming close to them.

There was still the choice of wood to decide. Many hives are now made of pine, but the traditional timber is cedar, a wood that is full of natural oils, weathers well, requires no treatment with chemical preservatives, and will not bend and buckle through the vagaries of English weather. Cedar hives would be more expensive than pine ones, but it was an expense I was happy to pay. I

still have those first hives and they are as sturdy as ever. I have tried pine hives since but always found them wanting; they warp and twist, and gaps appear which the bees have to close up. Yet my cedar hives, though not the beautiful red they were when new but now grey with age, still fit together with snug precision.

It was winter as I thought all this through, so I had plenty of time to get my hives in place before I could move bees into them in the spring. I scoured the Internet and perused catalogues to find what I needed and at the best price, placed my orders, and waited for the hives to arrive. What came to my front door were cardboard boxes full of dozens of pieces of wood, bags of nails, a few bits of metal, and some instructions. Wooden hives come as flat packs.

One of the constraints on the urban beekeeper is finding somewhere to work as well as somewhere to keep hives. I had to find space on my doorstep to hammer bits of wood together and my kitchen floor was covered in hive pieces. For many, the do-it-yourself aspect of beekeeping is one of the attractions, but as I pulled apart joints that I had fixed upside down, smashed a thumb nail with a heavy hammer, and swore in frustration as I struggled with parts, I quickly learnt my limitations. Next time I would pay someone else to put my hive together or spend the extra money on buying from a skilled craftsman. Certainly I would have a completed hive to hand to use as a reference.

Assembling your hive

It seemed more complicated than it was. In the end a beehive is no more than a nest of wild animals in a box. Or rather a series of boxes. At the bottom is a stand.

It is crucial that wherever you keep your hive it is on a firm and level base. An advantage for the urban beekeeper is that there are plenty of suitable places, from roofs and balconies to concrete terraces and hard standings. If you do site your hives on soil, put down a paving slab first and use a spirit level, or at least a good eye, to judge that it is level. Bees like to build their comb down vertically; a horizontal base will ensure that the hive walls and the frames inside are vertical, and the bees will be happy.

It is surprising just how many bits of wood you need as a beekeeper and like all the other urban beekeepers I know, I soon became adept at raiding skips for timber, off-cuts, discarded furniture, and so on, which I could recycle for my apiary. Old plastic milk crates are popular urban hive stands, but I thought I should have something made of wood for my public hives and so knocked together stands from old beams out of a house that was being gutted nearby.

The floor of the hive goes on the stand, a landing board at the front, not for the bees, who will happily fly in and out of the hive waiting in the air until

PLAN OF THE HIVE

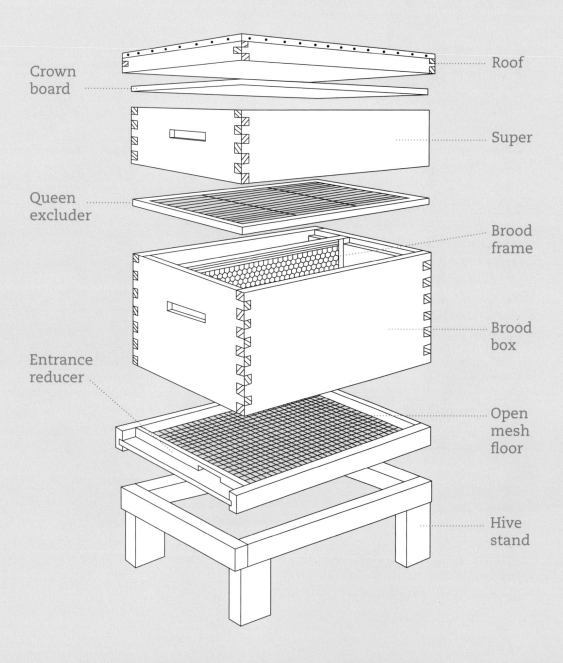

Crown board

Roof

Super

Queen excluder

Brood frame

Brood box

Entrance reducer

Open mesh floor

Hive stand

*National hive with homemade
stand for stability against wind*

they have an opportunity, but for the beekeeper to be able to watch the bees and to keep an eye open for pollen being brought back to the hive. Once upon a time, hive floors were solid, but the advent of the dreaded varroa mite, which can kill off a colony of bees once inside the hive, has seen the development of the 'open mesh floor'. The wire mesh sits in a frame. The bees cannot get through the mesh but the tiny mites can, and if they do fall through the mesh they cannot climb back into the hive. With that modification, the rest of the hive today is pretty much as Langstroth invented it.

On top of the floor sits the brood box. Between the front of the brood box and the front of the floor is an opening for the bees to fly in and out. It is the only opening in the hive. There should be no other gaps. The brood box is where the queen lives, where she lays her eggs and where the workers raise the brood—the young bees. It is filled with the moveable frames that Langstroth invented— different shapes and sizes in different styles of hives but all essentially rectangular frames within which the bees build their comb and which hang from supports at the top of the box, like files in a filing cabinet. And all separated from each other and the sides of the hive by that bee space of between ¼ and ⅜ of an inch (6 to 9 mm).

There are other boxes on top of the brood box

called supers. These are less deep than the brood box and are where the bees make and store their honey. To ensure that the queen does not lay eggs in these boxes, a sheet of perforated plastic or metal (queen excluder) is placed between the brood box and the supers with holes to allow the worker bees through but just small enough to confine the larger queen to the brood box below.

On top of the supers sits a crown board to close up the hive, and on top of everything sits a roof covered in metal or roofing felt to keep the rain out.

The hive today

Bees will live anywhere. They need space, darkness, and a tiny hole to come and go. So in towns and cities you will find wild and feral colonies in chimneys, roofs, church towers, voids in walls, and under floor boards. The beekeeper tries to offer bees a perfect home, the optimum living space, spacious enough for the colony, with plenty of room to store honey, dark, dry, and secure. For the beekeeper a good hive is one that can be opened,

examined, harvested, and husbanded with the least disruption to the colony within.

The modern hive, with its boxes piled one on the other, has ruled the beekeeping world since Langstroth invented it. But in recent years there has been a resurgence of interest in older traditional hives, based on the horizontal log hives of previous ages and of other cultures. These traditional hives survived in rural communities around the world. Now there are versions fashioned from wood that are ideally suited to the urban beekeeper. Instead of being encouraged to build upwards, the bees build outwards, the colony expanding lengthways in the hive, building their comb not on frames but on bars laid across the width of the top of the hive. Everything is contained in the one long box. When you raise the lid everything is in view and there is no need to lift out anything but one bar of comb at a time. A super laden with capped honey can easily weigh 35 pounds (16 kg) or more, depending on the type of hive. Shifting them can be back-breaking work. Not so the top bar hive, where the frames sit high off the ground and the beekeeper never has to lift the weight of more than a single comb of honey—though with great care as, without a frame to support it, the comb is fragile and easily broken. These top bar hives and a resurgence of interest in the Warré hive are part of the development of 'natural' beekeeping methods amongst those who want to keep their bees with the least possible human intervention. A futuristic top bar hive called the 'Beehaus' is the latest hive design to hit the market.

I had put together my first hives and they looked perfect for their setting. They were small, crucial in the heart of London where even in gardens space is at a premium, discreet while still being recognisably beehives, and with their cedar wood sides and metal roof, protected against the elements. I had even put a small set of railings around them to ward off young children and the sheep who spent their summers in the garden. Now all I needed was some bees to put inside them.

TOP TIPS

CHOOSING A HIVE

- Decide why you want to keep bees and what might be the most suitable hive.

- Check out the hives of other beekeepers and see what works best for you.

- Think about the amount of work involved with different hives and lifting the weight of a full box of honey.

- Imagine the hive in your space and consider shape and colour.

- Put down a paving slab to ensure a firm, level surface for the hive.

"Handling a box full of bees is an extraordinary thing to do, miraculous and scary—and heavy."

BUYING BEES

THE GOOD, THE BAD, AND THE SWARMY

The bees I needed were honey bees. There are many types of bee—bumble, carpenter, leafcutter, mason to name just a few—some twenty thousand in all and they are found everywhere in the world where there are plants. Only Antarctica is a bee-free zone. All the world's bees are vital for the pollination of plants, but there is one bee that has a very special place in our ecosystems and that is the European or western honey bee, *Apis mellifera*.

The honey bee is different from other bees in surviving through the winter months: for that it needs stores, and those stores are honey. The honey is made from nectar collected from flowers. The bees collect pollen, too, which they bring back to the hive as a high-protein food source to feed the brood, and as they forage from flower to flower they drop off pollen and so fertilise the plants they visit.

Other bees live in very small colonies which do not survive the winter—the bumble bee nests in colonies of just a few hundred and only the fertilised queen remains alive through the winter months, hibernating in the ground. Other bees

Hive on the roof of
a London pub

live solitary lives, laying eggs at the end of the summer which will not hatch until the spring. The honey bee lives in a large community, storing up honey in wax cells as a winter food supply.

The honey bee colony

A hive provides a home to a single colony of bees and that colony is built around a single queen. A strong colony in the middle of the summer might contain up to seventy thousand bees. A few hundred will be male bees, the drones, whose only function is to mate with queens from other hives and so propagate the species. All the rest will be workers, responsible for every duty in the hive, even feeding the drones. All the workers will be sisters or half-sisters, daughters of the queen.

To populate my two hives I needed to find two colonies of bees, one for each hive. Each colony would need a laying queen, one that had mated and was able to lay eggs, and thousands of workers to ensure the colony was strong enough to survive in its new home.

The two easiest ways to get hold of bees in Britain are by purchasing a nucleus or by catching a swarm.

A swarm is a colony's way of reproducing. If you think of a colony of bees as a super-organism, one giant creature, then that organism reproduces by splitting itself in half. The old queen leaves the colony with half of the bees while a new queen is raised by those left behind. The bees that leave the hive fly in a great cloud away from their old home and find a convenient place to rest. Usually this is on the branch of a tree, but urban bees can hang around in the most unlikely places. I have seen them on lamp-posts, shop fronts, and even on a motor scooter. They hang in a giant cluster, a bit like an American football or rugby ball, with the queen protected in the middle. Scout bees then go off in

search of a permanent new home. That is the time to capture the swarm and take it back to your hive.

You can trick the bees into thinking they have swarmed by taking the old queen and some frames of brood and stores out of the hive and putting them in a new one. This is called splitting the hive, the most common way in which new colonies are created in Britain.

Buying bees—swarms and nucs

There are about thirty recognised subspecies of *Apis mellifera*, that is, types of honey bee with distinct characteristics. With centuries of interbreeding, much of it as a result of human interference, there is also an endless number of cross-breeds—the mongrels of the honey bee world. Every subspecies and every breed has its own temperament. For the town and city beekeeper it is important as far as possible to find bees that have a temperament that is suited to the urban environment: bees that are gentle and healthy, not too 'swarmy', and used to an urban life. What I did not want for my first hives in the heart of the city with public all around, were bees that were aggressive and prone to swarming. And I did not want bees that were diseased. If that meant that they were lazy and would not make large amounts of honey, then that seemed a good trade-off to me.

The best thing to do when looking to buy bees is to find someone you trust and so know what it is you are buying. It can be a big investment. Swarms are a cheap, even free, way of getting bees, but the danger can be that you do not know what kind of bee you are getting. And if they have swarmed they are likely to be swarmy bees. The queen will be old, the colony may have disease, and they could be very bad-tempered. A bought nucleus on the other hand should be checked for disease and come with a newly mated queen and from a beekeeper who will tell you something of the colony's character. So for my first bees I went back to the beekeeper I trusted the most, my mentor John Chapple. John lives in a suburban semi in the west of London, a stone's throw from the motorway that leads to Heathrow. In the front of his house a couple of beehives stand sentinel—much better, he says, than dogs at protecting his property. At the back of his house the garden is a virtual bee Eden with hives everywhere, nucleus boxes ready for swarms, and, in the height of the season, even the occasional swarm hanging from a tree branch ready to be given a new home. John was happy to make up a couple of 'nucs' for me, and one afternoon he drove up to the museum with them in the back of his car.

John opened the tailgate and lifted out a box. 'You take this one,' he said as he passed it to me. Handling a box full of bees is an extraordinary thing to do, miraculous and scary—and heavy. The bees were very noisy, distressed and angry at being held prisoner.

We carried the boxes to the hive site and lifted the hives off their stands. Bees have a very precise sense of location and we wanted to put them exactly where their new homes were to be. And we did not want to transfer them from their travelling boxes just yet—they were far too angry for that. Best to let them settle down and get used to their new surroundings. We put the travelling

nucleus boxes on the hive stands, their entrances facing in the direction that the hive entrances would be, and opened up the fronts. A few scout bees flew out straight away, checking the lay of the land. There were some guard bees as well. Once word went around the colony that all was well, more bees came out, flying around to get their bearings, checking where they were, and learning the position of their new home. As long as they were far enough from their old home (three miles is the usual distance considered safe), the bees would not fly back to their old hive site, but learn instead their new location. We made a cup of tea and watched for a while. Soon pollen was being brought back to the colony and all seemed to be well. We decided that they had had quite enough stress for one day and would leave it until tomorrow to transfer them to their new hives.

Stings and how to avoid them

The following day we went back to the hives and put on our outfits. John with his years of experience is very relaxed about wearing protective clothing while working with the bees. He just puts on a veil, the most essential piece of clothing because the one place you really do not want to get stung is on the face, and the swelling from a sting to the mouth or eye can be especially dangerous. New to handling bees, I was much more cautious.

Bees rarely sting and will do so only if they feel threatened. Opening up their home can of course be perceived as a threat. The key, I quickly learnt, is to always work slowly and gently. Avoid sudden movements and move at the pace of the bees themselves. If a bee does sting it will die, leaving the sting, venom sac, and pump behind as it flies away. It will also leave a strong pheromone, warning the other bees of a threat, and one sting will quickly be followed by others. If you do get stung it is important to remove the sting, venom sac, and pump as quickly as possible, scraping them out with your hive tool so that no more venom is injected. And it is important to smoke strongly around the sting so that no other bees can smell the warning pheromone. Best of all is to avoid getting stung at all, and so a bee suit, boots, and gloves, all properly zipped up before opening the hives, should be worn on any hive inspection. This is especially true in small urban environments, where simply walking away from stinging bees is rarely an option.

Smokers and other equipment

I climbed into my bee suit, pulled on my boots, and lit the smoker. A smoker calms the bees. The smell of the smoke confuses their communication systems and, in an atavistic response from a time when all bees lived in trees, the colony thinks there has been a forest fire and retires into the brood box, gorging itself with honey on the way. It can be a while before a forest fire blows over and the bees prepare for a period sheltered from the passing flames. That at least is the theory. Once I had got the smoker going I zipped up my hood, put on my gloves, and smoked the front of the nucleus boxes.

We removed the boxes from the hive stands and replaced the hives.

BEE STINGS

When a honey bee stings, the sting, venom sac, and pump are left in the skin after the bee pulls away. Most of the venom will be injected in the first twenty seconds, but the pump can continue to inject venom for another two minutes. So it is important to get the sting out as quickly as possible to minimise the dose of venom. The best method is to scratch the sting and venom sac out, and a hive tool is perfect for this. Then smoke the area to mask the alarm pheromone in the sting and stop any more bees from stinging in the same area.

The vast majority of bee stings are harmless though they can for a few minutes cause a sharp pain, which then becomes a dull ache. Redness, swelling, and pain are all common reactions to a bee sting.

Keep your smoker going, close the hive gently, move away for a few minutes, and, once the sting is removed, apply a soothing balm, such as witch hazel or calamine lotion, to the affected area. When you are home, an ice pack or packet of frozen peas will help to reduce any pain or swelling.

In around one case in a thousand there is a more severe reaction. Severe reaction to bee stings is rare. There are between four and eight hundred admissions to hospital in England for all insect stings every year, the majority from wasps, which are far more aggressive than bees. There are on average two deaths a year in the UK from bee stings, compared to three from lightning strikes.

In the very rare case of an allergic reaction, symptoms can include a raised rash (weals), headache, fever, severe swelling of the face, tongue, and lips, and/or difficulty breathing. An allergic reaction will usually occur within one hour of a sting. Allergic reactions are more common after multiple stings.

Allergic reactions require emergency medical help if there is collapse and/or difficulty breathing.

Unfortunately even beekeepers that normally show little reaction to bee stings may react adversely the next time they are stung, so it is always wise to be prepared and ensure that help can be called in any emergency.

The bees' new homes would be exactly in the places they had become used to while flying the previous day. Then we opened the lids and very gently lifted out the frames of bees one at a time and transferred them to the new hives.

Our hive tools were essential. The ubiquitous hive tool is the Swiss army knife of the beekeeping world. Its sharp edge can get between the parts of a hive that the bees have stuck together with propolis. The tool can be used as a lever to raise one hive box from another or separate frames joined with wax or propolis. The tool's hooked end can lift out a frame from the others around it.

There were five frames in each of the nucleus boxes, and every frame was heavy with bees, brood, and stores. The hives had room for eleven frames, so we put three empty frames at either side of the brood box and placed the bee-covered frames from the nucleus box between them. We worked methodically and slowly, making sure the frames were in the same order in the new hives as they had been in the travelling boxes. And we looked out for the queen as we did so. The queen is not always easy to spot and she will do her best to hide from you, keeping out of the light if she possibly can. Many beekeepers mark their queens with a small blob of paint on her thorax. This makes her much easier to find. There is an internationally agreed colour code for queen marking so that you can tell her age by the colour of the blob of paint on her.

Once the queen had been rehoused, the rest of the bees that were not on the frames or out flying would follow into the new hive. After we had moved the frames, there were still bees left in the bottom of the boxes. We upended each over the hives and gave them a sharp tap with our hive tools. Most of the remaining bees dropped off and joined their sisters. The ones that did not, we left to walk in through hive entrances as we put the almost empty nucleus boxes against the landing boards.

The bees seemed hardly to notice what we were doing and as we replaced the hive roofs a few minutes later they were behaving as if nothing had happened, flying in and out for pollen and nectar and no doubt busy inside the hive building comb onto all the

HIVING A NUCLEUS

1 Nuc boxes on hive stands

2 Nuc box and prepared hive

3 Lifting nuc box from stand

4 Lifting lid of nuc box

5 Frame into hive

6 Using hive tool

new frames. They were indeed gentle and ideal for this most public of inner city settings.

Packaged bees

In the United States the easiest and most common way to obtain your bees is to buy a 'package'. Instead of being on frames with brood and stores, the bees come in a mesh-sided box and are sold by weight, two or three pounds in a box. The queens are kept separate, in their own little cages the size of a matchbox with some sugar and a few attendant workers. When the packages arrive, instead of lifting frames of bees into the hive, it is a simple matter to tip the bees into your waiting hives and then gently place the frames in amongst them. You have a hive full of bees in an instant but no brood and no stores. The bees need to build comb, bring in pollen, and make honey. And the queen needs to lay as soon as she has been accepted into the colony and there are cells for her to lay eggs in. Between the lid of the queen cage and the queen with her attendant bees, is a lump of sugar. The lid of the queen cage is removed and the cage is suspended between a couple of the frames. The bees will take a while to adjust to the pheromones of their new queen but by the time they have eaten through the sugar and released her, she should be accepted into the hive. Bees have been sold in packages in the United States since Langstroth first invented the moveable frame hive, though the bees that Langstroth put in his hive were not indigenous American bees. The honey bee is not an American native.

The honey bee comes to the USA

It is thought that *Apis mellifera*, like *Homo sapiens*, first appeared in the south of the African continent and spread across the world from there. While man went everywhere, *A. mellifera* confined itself to Africa and Europe. As Europeans colonised the rest of the world, they took the honey bee with them.

The first hives were brought to North America by the early settlers. If I find

WHICH YEAR REARED GREAT BEES

There is an internationally agreed colour code for queen marking so that you can tell her age by the colour of the blob of paint on her. Remember it with this handy mnemonic.

White for years that end in 1 or 6. *Yellow for years that end in 2 or 7.* *Red for years that end in 3 or 8.* *Green for years that end in 4 or 9.* *Blue for years that end in 5 or 0.*

it scary driving boxes of bees across London, how much more daunting a task must it have been to transport entire hives across the Atlantic on sixteenth-century sailing ships? Those first hives of bees would have been valuable commodities. To this day American bees retain a different status in law to their English cousins. No one can own bees in England; they remain wild creatures, *ferae naturae*, in the custody of their keeper, hence bee*keeping*. In the United States, bees can be owned and are regarded in law as chattels—moveable possessions.

When John Eliot translated the New Testament into an indigenous American tongue in 1661 he found that there were no words for 'wax' and 'honey' in the language. The native Americans called the new insect 'the white man's fly'.

On 5 December 1621 the Council of the Virginia Company in London sent a letter to the Governor and Council in Virginia with the following news: 'We have by this ship *The Discovery* sent you divers sorts of seeds, and fruit trees, as also pigeons, connies [rabbits], mastiffs, and beehives, as you shall by the invoice perceive; the preservation & increase whereof we respond unto you.' The ships and their bees arrived in Virginia in March 1622.

There are various stories and reports of how bees were transported on board ship for the six to eight weeks of the voyage to the New World. It seems likely that most were in skeps. According to one account, 'The bees were placed on deck as follows: A strong oak platform was built on the stern of the ship, the crate

containing the skeps was bolted to the platform facing the sea at the rear of the ship. This kept the bees as far as possible from the ship's crew and passengers.'

The bees were sometimes left to fly, navigating we must assume only within the confines of the deck of the boat, because the position of their home was constantly changing in relation to the sun. Sometimes the hives were packed with ice to keep the bees clustered together inside. There must have been many losses. By all accounts the bees that did survive the crossings flourished, and within a hundred years there were feral colonies throughout the eastern seaboard to be hunted, caught, and hived.

It was in 1853, over two centuries

Left
Package of bees arrives

Above
Last bees leave the package

after they first came to the east coast of America, that honey bees arrived in California. Twelve hives were purchased in Panama, transported across the isthmus and then carried aboard ship to San Francisco. Only one hive survived the trip, but once there it flourished and eventually produced a number of swarms.

Before long, farms grew up with imported European fruits to be pollinated by the imported European honey bee.

Today vast agricultural industries are dependent on the bees, and thousands of hives are trucked around the continent to pollinate various crops as they come into season.

There are plenty of other, native, bees in America as there are other native bees around the world, and all play their part in pollination. Other plants, especially native plants, require native bees adapted to their pollination needs, and the keen gardener or ecologist can provide homes for them, too. You do not have to keep a hive of honey bees to be a beekeeper, and any gardener should consider providing a home for their local native bees.

Moving bees and beehives

Since installing my first hives, I have transported bees all over England and even brought them in from overseas. There have been some adventures along the way. It is always best to move bees over night. You can close up hives or travelling boxes at dusk when all the flying bees are home, so that none are lost, and a night-time drive across London with a box of bees is the best of bee drives. The bees are quiet in the box, at least to begin with, and with little traffic about, trips can be quick. Such drives are not always without incident. I delivered hives late one night to a roof in Carnaby Street, only to find a crowd of revellers celebrating the launch of an art show when I drew up in Soho. The art lovers surrounded the car, mistaking me and my colleague for performance artists as we clambered out in our bee suits and lifted the buzzing boxes of bees from the boot.

Moving bees across the city during the day is always more stressful than doing it at after dark, to the bees and to the beekeeper. The traffic seems to move at a snail's pace, the bees get ever more angry at their confinement, and they express their anger with ever more noise. I always take my kit with me just in case something goes awry—suit, smoker, and, most important of all, gaffer tape. I was driving bees across London one morning, having picked bees up in boxes from the far south of the city. As I sat in traffic at the Elephant and Castle, I glanced in my rear-view mirror and noticed bees flying about in the back of the car. The only thing to do was to pull over to the side of the road, don suit and smoker, and check where the bees were coming from. One of the boxes had not been securely closed, and bees were coming out to scout the situation. As the rush-hour commuters watched me from their cars, I sealed up the entrance with tape and set off again. I had lost a few flying bees and still had one or two loose in the car, but at least I had averted commuter chaos. Now I

always check that travelling boxes and hives are fully closed up before I move them. It might not matter much if your bees escape in a van in the countryside, but it is not what you want to happen in a car on a major interchange of roads in the middle of the city.

There is now a worldwide trade in bees and especially in queens. I have had bees delivered by Royal Mail, though they were a sorry sight having 'baked' in transit and the comb on their frames melted into a sticky mess of honey, dead bees, and wax. Luckily the queen was still alive as were half of the bees, and a quick rescue mission enabled them to recover. Transporting bees on frames of comb through the post or by courier is fraught with hazard. The bees get hot, the wax melts, the brood die, and even the queen may not survive despite all the best efforts of her workers. If you transport bees on frames in a nucleus or travelling box, do all you can to keep them cool. Make sure you give the boxes as much ventilation as possible and always have the length of the frames facing the direction of travel so that as you stop and start they do not knock into each other.

Packages of bees seem to travel well.

There is nothing but bees in a package, no frames, no honey, no wax or brood. Nothing to melt. Just thousands of bees and a queen safe in her little plastic cage. Thousands of packages are airlifted around the world every year. There is a trade in mated queens too. If you replace an old or failing queen with a new one, you should know her temperament and how well she has mated. A docile queen introduced into a vicious hive will have replaced the existing troublesome bees with her own docile young within a few weeks. I have bought newly mated queens flown in from New Zealand. I have dealt with Slovenia beekeepers exporting the wonderful *Apis mellifera carnica*, the Carniolan honey bee famed for its gentleness and, though a native of the mountains, an ideal urban bee. I have

Left and below
Hives at the Lancaster London Hotel

even come across American 'boutique' bee breeders who claim to be able to provide the perfect strain of bee for every beekeeping need.

In the end, though, I have come to believe that whenever we can we should source local bees from local beekeepers, bees born and bred in the city and ready for an urban life.

BUYING BEES

- Buy from someone you know and trust.

- Choose bees to suit your needs.

- Buy local.

- Have the queen marked before you buy.

"Think of it as a house that you are about to take apart room by room, while most of the occupants are at home."

LIFE INSIDE THE HIVE

LOOKING AFTER YOUR COLONY OF URBAN BEES

For bees the year starts in the spring. Through the winter months the colony of thousands of workers and a single queen will have been quietly clustered together. The queen is the only bee able to mate and lay fertilised eggs. She is vital to the survival of the colony and will be protected by the workers, all female and all her daughters, through the winter months, surviving on the honey they have stored during the previous summer. Once the days begin to lengthen and the temperature rises, the bees will start to venture out from the hive searching for food—protein-rich pollen and sugar-rich nectar.

Pollen is the fine powder in flowers that contains the plant's sperm cells; as the bees collect pollen from flower to flower, some will brush off and so pollinate the flowers. If there is pollen to be found, the bees will bring it back to the hive as forage, and with food in the colony the queen will begin to lay eggs. Three days after the eggs have been laid they will hatch, and the new

Apiary at the Lancaster
London Hotel

A frame of bees

larvae will need to be fed by the adult bees with the pollen.

If all goes well, the weather remains warm, and sources of pollen continue to be found, the queen can lay up to two thousand eggs a day. Lots of pollen is needed to raise the brood, the larvae that hatch from the eggs, for the six days until they are sealed in their cells to pupate, and lots of energy-rich nectar, the sugar-rich liquid that flowers produce to attract the bees, is needed to make the honey which both adults and brood need to feed on. Quickly the colony expands from five or ten thousand winter bees to upwards of seventy thousand in the height of summer. The frames in the hive will soon be full of eggs, brood, pollen, nectar, and honey.

You can see all this when you inspect a hive. There are many reasons to open up the hive and inspect the colony. You want to know that all is well. That there is a queen and that she is laying. That the bees have enough room as they expand for the queen to lay eggs and for the workers to make and store honey. That there is no disease. And for many beekeepers, there is the simple delight of seeing the bees at work.

Best time to examine the hive is the middle of a warm, sunny day. Many of the older flying bees are out, foraging for pollen and nectar, so you have mostly the house bees to deal with. It is hot inside the hive, the bees keeping their home at a constant temperature of 36°C. A rush of cold air in as you open it will chill the bees and the brood. Too cold and the brood will die, and even on a hot day the bees will have to work hard to get the temperature back up once you have finished your inspection.

But as with all beekeeping there are other approaches. I have even come across night-time beekeepers, inspecting their hives after a long day at work when the sun has set, and the bees, disinclined to fly after dusk, remain placid as the hives are opened.

Do not rush. You are about to open up the bees' home and flood with sunlight a place that is always pitch black. Think of it as a house that you are about to take apart room by room, while most of the occupants are at home. The bees will be no keener on your taking off their roof than you would be if someone came along and removed the roof from your home. Start by smoking and waiting for a few minutes. Then begin to gently and quietly take the hive apart, making sure that you keep everything in order so you can put it back the way you found it.

Essential kit

First, though, you need to put on your protective clothing. There are many types of bee suit. Some people just use a veil, others a tunic with veil attached, others a full suit covering all the body. Some veils are like ones used in fencing and fall easily away from the face. Others are circular. It can be as baffling to choose your bee suit as it can be to choose your hive. And it can be expensive; best, if possible, to try a few different ones on and find what feels most comfortable for you. You will spend a lot of time in your bee suit and always when the weather is at its warmest. There are a few guiding principles. First the material should be smooth, nothing that the bees can get caught on. Second the veil should be well away from the face. A sting to the mouth, eye, or ear can be a worry. Third the suit should be roomy. Better a baggy suit far too big than one that stretches across your tummy or is so tight that it restricts your movement. As for colour, to paraphrase Benedick in *Much Ado About Nothing*, it can be of whatever colour God pleases, though a light colour, especially the traditional white, will keep you cool in the sun and show up any bees that are on you when you come to take it off.

For quick inspections I take a tunic or just a veil with me, but if I am taking a hive apart, transferring a colony, or doing inspections on hives I have not seen before, I take every precaution and wear a full suit with Wellington boots. I have grown more cautious over the years, perhaps true of everything in my life, and you never know the temperament of the bees in a strange hive. I even wear gloves. Gloves are hotly debated amongst beekeepers, especially the older ones. Gloves (they say) spread disease, and they are cumbersome and limit your ability to manipulate the hive. All that may be true. It is also true that they can stop you getting stung. Sheepskin gloves are the classic but impossible to wash and

difficult to clean except with saddle soap. Plastic gauntlets, which are washable, are available, and ordinary kitchen gloves can provide some protection, are disposable, and will keep your human smell away from the bees.

Do not forget your feet. Bees will not always fly, and bees that fall from the frames as you open the hive will often trudge home. Or climb up your leg. Wellington boots are a good idea.

In the end, wear whatever makes you feel both comfortable and protected. Have a last check to make sure all your zips are fastened, and before you pull on your hood, light your smoker: you do not want the flames to catch light to your veil.

Lighting a smoker is the most difficult job in beekeeping. For the urban beekeeper extinguishing the smoker is the other most difficult job. You can use all manner of fuel, from commercially made products to old egg boxes and dried leaves. A pleasant smell is nice for both bee and beekeeper, and my favourite fuel when I can get it is dried sage. I brought some Native American 'smudging sticks', little bound bundles of sage, back from a trip to New Mexico and they proved to

Kitting up   *The author in full bee suit*   *Bee-proof boots*

be the best fuel I have ever used, easy to light, aromatic to smell, and once lit would keep going with little or no attention while I worked on the hives. The sage burnt with a cool gentle smoke, just the thing for puffing into the hive and calming the bees. Only when the hive had been closed up and I had to extinguish the smoker did it present a problem. At the Natural History Museum it was easy. Like any rural beekeeper, I would pull two or three leaves off a tree nearby, roll them up and stuff them in the top of the smoker, leave it on the ground, and go and make a cup of tea.

Eventually the lack of air would cause the smouldering fire to go out.

It was a different matter when working on the hives on the roofs of the Ted Baker building and the London School of Economics. The smokers there had to be brought inside lest they fell, rolled, or were blown off the roof. With detectors and alarms everywhere, they had to be without a whiff of smoke left in them, so there we became very careful. We used no more fuel than we thought we would need, always had something to hand to block the smoker (there are no trees on those roofs), and had a metal

Checking the smoker

Smoking the hive entrance

box with a tightly fitting lid to store the smoker in between visits. There are some people who prefer not to use a smoker at all, and instead will use a fine mist of water from a spray can to calm the bees. The dampness keeps the bees on the frames as you examine them and discourages them from flying.

Finding the queen

Suit on, smoker lit, and hive tool to hand, you are now ready to inspect your hive. Always work from behind the hive.

News will have got out that something is happening to the colony as soon as they smell you or the smoker, and bees that are out and flying will soon be returning home. You do not want to be standing in their way as they try to get in the front of the hive.

Lift off the roof and lay it upside down to one side. Then you can put all the other hive parts on it. Any bees that fall off the boxes as you remove them and are unable to fly will be safe in the upturned roof. Next will be the crown board. In the middle of the season when the bees are at their most numerous

*Smoking down through
the queen excluder*

and they have stored lots of honey, there will be heavy supers to lift off. With a nucleus or overwintered colony early in the season there should be no supers to remove. Once any supers have been put to one side, the next thing you will need to take off will be the queen excluder. Ease it off carefully and shake any bees from it back into the brood box. Smoke as you go.

Keep an eye out now for the queen. You do not want to lose her. It is easily done as she can be difficult to spot even when marked. The queen is the most light sensitive of all the bees in the hive. She will leave only twice in her lifetime, once to mate and again to swarm. Apart from those few days, hours maybe, she will spend her entire life in pitch darkness. Even as you remove a frame, if the queen is on it she will be remarkably quick at moving to the side away from the direct sunlight. Nikki Vane keeps her hives high up on a roof terrace in Bermondsey. One afternoon I was helping her inspect her hive. It was a beautifully sunny day and we assumed the queen to be deep in the hive. As we searched the frames for her she was nowhere to be seen. No matter, we knew she was there somewhere. There were brood and newly laid eggs in the hive, and even if we had not seen that, the pollen being brought in by the foraging bees was a sure sign that there were newly hatched young to be fed. As we closed up the hive, there was the queen sitting on top of the queen excluder as we replaced it on the brood box. She must have been there all along: we had unthinkingly tipped the slotted metal sheet upside down as we removed it. Luckily for us, and for the colony, we saw her. I have known at least one beekeeper who has had a queen fall from a frame as they have removed it from the hive and then trodden on her before realising.

Working through the hive

Working from one side of the hive, you can remove each frame, inspect it, and then return it. To allow yourself and the bees space to remove the frames without 'rolling' the bees on each other, it is wise to take out the first frame, shake off the bees, and put it to one side. Some

*Opening a hive on the Ted
Baker building in London*

*Working through
the frames*

beekeepers like to keep a plain piece of wood in place of the end frame. This is called a dummy board. It will have no bees on it and so when removed allow you the space to move the frames without having to leave one outside of the hive. If you do have to leave a frame out of the hive, best to leave it propped against the entrance so any bees on it can easily return to the colony. But check the queen is not on it first.

Some people find it is easier to lift frames using not the hooked end of their hive tool but a frame grip, a tool specially made for the purpose. In examining the frames there is another invention which, bought or made at home, can be of real benefit to the urban beekeeper. It is called, like something from a box of magic tricks, a manipulation cloth. It is a rectangular metal frame, just longer and wider than the top bar of a hive frame. Attached to the frame on each long side are lengths of cloth. Placed on the top of an open brood box, the manipulation cloth allows you to look at or remove one frame at a time without all the bees flying up from the other frames. The result—fewer angry bees flying around, which in a tight urban space can be a real advantage.

The outer frames may have little on them early in the season, just cells being drawn out with new wax. As you get further to the centre of the brood box you will find cells full of nectar of varying

Left
Dummy board

Below
Moulded aluminium frame grip

Right
Young beekeepers checking the bees

colours if the bees have found a variety of forage. Some of this nectar will have been converted into honey and the bees will have sealed it into the cells with a thin white cap of wax.

Next you will find pollen and eventually brood: tiny eggs in some cells; glistening, white pupae in others; and larvae sealed with a biscuit-like capping in others.

Closing up

If all is well, you can put the hive back together, and the colony can be left to its own devices until your next visit. Make sure everything goes back just as it was. The frames should all be in the right order, the queen excluder on top of the brood box, any supers with the frames running in the same direction as the ones below, then the crown board, and finally the roof.

There will have been bees flying everywhere while you were working on their hive, but once closed up they will quickly return inside and things will quieten down. A few guard bees will remain outside, actively defending the colony. Keep an eye out for them as you take off your bee suit; and if you have

someone with you as a bee buddy, check each other for bees before you take off your suit and go back inside. It is the moment I seem most often to get stung. Oh, and don't forget to extinguish your smoker.

INSPECTING YOUR BEES

- Choose a warm sunny day to open the hive.

- Wait till the bees are flying.

- Decide what you are looking for before you begin.

- Get your smoker going (and spend time practicing this).

- Be slow, calm, and gentle.

- Check the queen is laying.

- Make sure the bees have plenty of room in the hive.

- Make sure they have sufficient stores until the next inspection.

- Keep an eye out for disease.

- Ensure everything is back in place when you close the hive.

"All this happens on the wing."

SEX AND THE CITY BEE

THE LIFE CYCLE OF A HEALTHY COLONY

For me, keeping bees has become a perfect hobby, a quiet and meditative escape from a hectic life in the theatre world. It has made me more in tune with the weather and the seasons, aware of the days shortening and lengthening, and alert to flowers and blossoms that I had passed without thought for years. I have learnt what goes on in the hive and how to observe it, and as I have taken on more hives my life has begun to adjust to the beekeeping year. Some years it seems as if spring will never come, and I worry that my bees will not survive until the warm weather returns. Spring of course always does come eventually, and in London it comes earlier than in the rest of the country. One year, a hive maker from the north east of England delivered me a couple of new hives that I was desperate for. I had colonies getting ready to swarm and needed them to split them into additional hives. He did not believe my urgency. There was no sign of spring where he lived and certainly no sign of colony expansion amongst his bees. It was only when he drove to

Collecting nectar

London that he realised just how far advanced we were down south.

The start of the season

On a bright sunny morning you will see a few bees venturing out of the hive and returning with little bundles of colour on their back legs. It will be pollen being brought back into the hive to feed the young. The queen has started to lay and the bee season has begun.

That small cluster of bees that have huddled together through the winter on just a few frames is about to rapidly expand to fill the hive. The combination of warm weather and forage being brought back into the hive stimulates the queen and she starts laying eggs. The winter bees that have kept the colony alive through the cold dark months will soon die off, their work done, as new bees are born. It takes twenty-one days from an egg being laid to a new worker bee emerging. So between the queen beginning to lay and the colony expanding, there is a three-week lull. The colony may even decline during that time as the winter bees die off. The eggs hatch as larvae in three days. They will need feeding with the pollen from early blossoms. Six days later the larvae are sealed in their cells and take a further twelve days to pupate, eventually emerging as bees. Each new bee will clean up the cell it was born in and make it ready for the queen to lay another egg. The cycle will continue through spring and into the summer. As the new bees emerge, the expansion of the colony can be rapid, a thousand or more bees a day. The five or six thousand bees that have come through the winter and nursed their successors are replaced by the tens of thousands of bees who will become the colony during the summer. All will have work to do.

It has taken three weeks for an egg to become a new worker bee. For the next three weeks, that bee will be busy in the hive, doing all the jobs that need to be done—cleaning, making honey, feeding the new larvae, tending the queen, and taking out the dead. The last of the winter bees have done their work and die exhausted outside of the hive. Those that die in the hive will be swiftly

Left
Hive inspection at the start of the season

Below
Inspecting a frame from hives in Penzance, Cornwall

removed. After three weeks' working in the hive, the new bees are out flying and will be foraging for the pollen that the last of the winter bees had been bringing in. The summer bees, unlike their winter sisters, have a much briefer and busier life. They have just another three weeks to live, weeks they spend as flying bees, foraging, (some for nectar, some for pollen), guarding the hive entrance, scouting for food.

There are few more enchanting sights than watching your bees bringing pollen back into the hive. The pollen is stored in 'bags' or 'sacs', one on each of the back legs of the bee. The bee compacts the pollen as it is collected, rubbing it together between her legs so as to be able to get as much in the bags as possible. The pollen is clearly visible as colourful bundles on the legs of returning bees. There can be 15 mg of pollen carried back by each bee—a sixth of her body weight. If you have a landing board on the front of your hive, you can watch as the bees sometimes wait for their chance to get back into the colony. Others will wait on the wing, and some will fly straight in without having to wait at all. On a busy day the foraging bees will dart in and out of the hive, travelling at over 20 miles (30 km) an hour. Every colour tells a story of where the bees have been foraging, and soon you will distinguish between the subtlest of shades.

Within weeks there will be fifty or sixty thousand bees in the hive and all will be busy. The beekeeper has to be busy too to keep ahead of them. An expanding colony needs space. Space for the brood and space for stores. If you start with a nucleus, the empty frames at either side of the nucleus frames will fill with brood, pollen, and some honey stores. An overwintered colony will expand in the same way. Confined in the brood box at the bottom of the hive, the queen will be unable to get through the queen excluder and into the super boxes above. The workers will use the space in the supers to make and store honey. They do this, not for us to harvest, but to provide stores to get through the coming winter for which they are planning from the beginning of spring. A full colony can build comb and fill it with honey very quickly. In a week a super can be heavy with stores and a new one needs

to be put on top of it. If you do not keep a step ahead of your bees, they will have nowhere for their stores and you will be in trouble. Have your supers ready and full of frames, and put them on sooner rather than later.

On hot afternoons there may be small clouds of bees around the hive; these are newly flying bees on orientation flights, learning the location of their home before they fly off and forage. There may even be a cluster of bees on the front of the hive, cooling down outside from the heat of the colony within.

Frames, foundation, and wild comb

Honey bees live on combs of small hexagonal cells made of wax. The wax is made inside the bees and extruded as tiny sheets from their abdomens. The wax is fashioned into a perfect matrix of cells, and it is in these cells that the life of the hive takes place. In the wild, bees will build their comb down from the top of the space they have chosen for their home in two layers, with cells on either

side of each sheet of comb. Between each sheet is the optimum space that the bees need to move about and get on with their work without disturbing each other. In the wild the comb is built down in long looping shapes, the biggest sheets of comb in the middle, the smaller on the outside. Inside a hive, the beekeeper places frames for the bees to build their comb on, so that the comb

and the colony can be taken apart and reassembled. Often a thin sheet of wax with a pattern of hexagonal cells on it is placed in each frame to encourage comb building and to give the bees a head start.

Some beekeepers prefer to use empty frames with no more than a small starter strip at the top of each to get the bees going. If you do not put frames in your hive, or leave a frameless box on top of another, the bees will very quickly fill it with wild comb, which they will build down from the top of the hive until it joins with the boxes below. One summer I failed to keep ahead of my bees at the museum. They were expanding rapidly, and I put a new super full of frames in one of the hives and left an empty super above it, intending to come back the following day having made up the frames to put in it. But other hives distracted me, and it was a week before I found time to return to the museum. I lifted the lid of the hive and a mass of wild comb and honey fell off. The bees had not only filled the super of frames but had filled the empty super above it. Wax, bees, and honey were everywhere and I had learnt a lesson— always be a step ahead of your bees and never leave a space without frames, or the bees will fill it with wild comb.

Swarms and swarming

Bees swarm, and though it's not one you want to see too often in an urban setting, a swarm can be a spectacular sight. Half the colony leaves, with the queen, and begins the search for a new home. That search is carried out by scout bees that go off looking for a suitable space—dark, dry, easily protected, and big enough to house a thriving colony. They will report back to the rest of the swarm, who will be clustered with the queen in a big ball of bees, often hanging from the branch of a tree close to the original hive. In the city a swarm will find all sorts of places to wait for their new home.

I was travelling back through Soho one hot afternoon when I noticed a swarm clinging to a lamp-post. Oh dear, I thought, these must be something to do with me. By the time I returned with an experienced beekeeper and a cardboard box to collect the swarm in, another beekeeper had already scraped most of them into another box. People were dashing past on their busy city lives as if nothing was happening as bees flew around. I was looking after two hives nearby at National Magazines House, and once things had been dealt with went around to check them out. All was well. It transpired that the bees were a feral colony, disturbed by workmen knocking down a wall.

LIFE CYCLE OF THE HONEY BEE

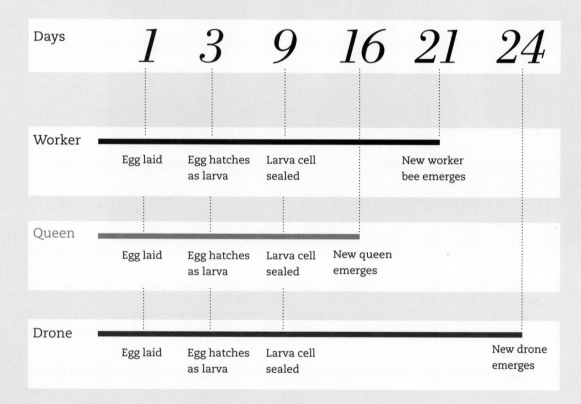

Days

| | 1 | 3 | 9 | 16 | 21 | 24 |

Worker

Egg laid | Egg hatches as larva | Larva cell sealed | New worker bee emerges

Queen

Egg laid | Egg hatches as larva | Larva cell sealed | New queen emerges

Drone

Egg laid | Egg hatches as larva | Larva cell sealed | New drone emerges

Swarming bees are at their most docile. They will gorge themselves with honey before they leave the hive, preparing for days without a home and with no stores to live off. As they huddle together, often not far from the hive they have left, they can be scooped up, put in a box, and taken away. Usually they will be somewhere more conducive than a lamp-post, most often on a tree or bush where the branch covered in bees can be snipped off and the cluster of bees knocked off in one quick movement.

In another part of town, at the Lancaster London Hotel, one of the hives had swarmed and the swarm was hanging in a cluster on the branch of a tree over the nearby Bayswater Road. One of the hotel's 'Bee Team', who worked in the engineering department, was keen to get the cherry picker out and collect them, but they were high above the road and beyond our reach. All we could do was keep an eye on them. We watched the bees hanging over the city traffic as day followed day. The weather turned cold and wet, yet still they hung on. They were in no rush to make their choice of a

new home. Then one morning they were gone. We never knew where.

The swarming instinct is common to all honey bees, though some are more prone to it than others. Outside of towns, swarming can often go unnoticed, but swarming by urban bees can cause all sorts of problems. I have been called out to the BBC Television Centre, where a swarm had made its temporary home on a satellite dish, and to London City Airport, where a swarm was clustered on a runway light and disrupting flights. Urban beekeepers, in choosing their bees, are always looking for ones that are not just gentle but also as little prone to swarming as possible.

Swarm control

The swarming season begins when the hives have built up to full strength and when there is plenty of forage around for the bees to build and populate new hives with. It is the busiest time of year for the beekeeper and not the best of times to take a holiday. Before bees swarm they

Wild comb hanging from frame

will have begun making a new queen, for the half of the colony to be left behind, in a special queen cell. A weekly check on the hives is essential during the swarming season so that you can keep an eye on any queen cells and decide what to do before they are sealed. Best to avoid an emergency swarming situation if you can.

Something I noticed at two of my London apiary sites, one the roof of the London College of Fashion with some modest container planting around and the other a vast estate in Pimlico with an award-winning expanse of gardens, is that swarms of bees seem to love *Photinia* and will always choose to cluster on

this bright red 'Christmas berry', which flowers just as the swarming season begins. Easy to grow in sun or partial shade, it has become my lure plant, keeping swarms close to their home hives and so easy to capture.

One of the many good reasons to start with a small nucleus is that swarming is unlikely to be an issue in your first year. By your second season you should be comfortable enough with your bees and knowledgeable enough about what is happening in the hive to be ready for anything. There are various ways of controlling swarming by making an 'artificial' swarm—splitting the hive by removing the queen with frames of bees

Collecting a swarm in Wellington Street, Covent Garden, 1881

SWARM OF BEES IN WELLINGTON-STREET, STRAND.

from one hive and putting them into another, often a nucleus box, so that the bees think that they have swarmed all on their own.

If you do split your hive, you now have an extra colony of bees to keep for yourself or pass on to another, perhaps novice, beekeeper just as someone passed one on to you to get you started.

Virgin queens

Meanwhile, back in the hive that the swarming bees have left behind, the rest of the colony will be waiting for the emergence of a new queen. Before the old queen left the hive, she will have laid eggs in special cells often built on the bottom of the brood frames. These queen cups, as they are called, are big enough for the hatched eggs to swim in a bath of royal jelly that is their exclusive source of food. Excreted from two glands on the heads of the workers, this special food kicks the reproductive systems of the developing larvae into play and they become fertile queens, not sterile workers. Once the old queen has swarmed, new queens will emerge. The first new queen to hatch will destroy the other queen cells around her. If more than one new queen hatches at the same time, they will fight to the death, stinging each other until only one survives. Unlike workers who die when they sting, queens can sting again and again without harming themselves. The victor, a virgin queen, will then be ready for sex.

*Collecting a swarm in
Hong Kong*

The queen mates

The new virgin queen will need to mate if the colony is to survive and for that she needs male bees from other hives. These are called drones and there are always a few in the hive, hanging around waiting for sex. Every once in a while they will leave the hive and congregate together, drones from all the hives in the area, usually at a high open place, and fly around waiting for the queens to arrive. Each queen will have sex with as many males as she can, and every male will die as it mates. All this happens on the wing. Once she has mated successfully, full with enough sperm to last a lifetime, the queen will return to the hive not to leave again until she too goes off with a swarm just as her predecessor did.

And so the season continues. With its new queen, the colony, now half the size it was before the swarm left, will have to build up its numbers and produce honey stores for the coming winter.

As the pollen disappears and the nectar dries up, and as the days get shorter and colder, the beekeeping season begins to draw to an end. It is always earlier than you expect, and though it might still feel like summer to you, for the bees the year is already coming to a close. The queen will reduce her egg laying, and those new bees that do emerge will be winter bees, bred to survive

*Swarm on photinia
at London College of
Fashion*

through the long cold months ahead and rarely venturing from the hive. Their work done, summer bees will start to die away. Drones, of no use now that there are no virgin queens to be fertilised, are chucked out of the hive by their sisters and left to perish, cold and hungry, in the outside world. It's schadenfreudian to watch them pathetically trying to get back into the hives only to be rebuffed by guard bees at the entrance.

The bees have survived through the season and have stored plenty of honey to feed on over the coming winter. It is harvest time.

TOP TIPS

SEX AND THE CITY BEE

- Inspect once a week.

- Check the bees have enough room to expand.

- Look out for signs of imminent swarming.

- Keep an eye out for disease.

- Make a record of when you visit and what you see.

- Try to keep a step ahead of the bees and have frames and boxes ready.

- Seek advice when you need it.

"If you have dandelions on your lawn, try to leave them for the bees."

THE NECTAR GARDEN

FROM PLANTS TO HONEY

There is a pussy willow (*Salix caprea*) tree outside my front door. It is not very big, perhaps five feet tall in its pot, and as there is not much space outside my front door, I brush against it every time I go in and out of the house. It is the most useful of all my container plantings, for my pussy willow tells me when the beekeeping season is starting. The catkins are amongst the first flowers of the year and a harbinger of spring. When the vivid yellow pollen of the salix stains my jacket as I put my key in the door, I know that the first pollen of the year is ready to be taken into the hives.

No year is the same, and the beekeeper, like the bee, has to respond to the ever changing seasons and be alert for their coming. Plants are a much better indicator of the turning of the seasons than weather forecasters or calendars. You can have a plan of what needs to be done at the hive in which months, but only an eye on the blossoms will give you a true guide. One thing is certain: as an urban beekeeper you can be sure that there will be pollen earlier in the year than outside of town.

The White House beehive

Left
Foraging on a dandelion

Right
Foraging on thyme

The days can still be very cold when the salix is in bloom, and sometimes on my first visit to the hive, I will not witness foraging bees coming in and out. But if I see bees bringing pollen in on a sunny day, I know that the queen is beginning to lay. Another way to tell if pollen is being brought into the hive is to pull out the tray under the mesh floor, as some of the pollen will have fallen through and collected onto the tray. A sprinkling of pollen signals that all is well in the colony. Often, the salix will be there along with other pollens of early spring, such as crocus and snowdrops, their pollens a vivid orange; the brown pollen of gorse and the khaki of hazel come a little later.

If your hive is situated near a window of your home or office, you can keep a close eye on the activity of the bees in comfort. If you sit and watch the pollen being brought into the hive, the bees laden with heavy sacs, it is proof positive that the queen is actively laying and there is brood to feed.

The winter bees feed the next generation with pollen, nectar (or diluted honey if there is no nectar around), and a little bit of royal jelly extruded from glands on the heads of the nursing workers. Each newly hatched egg needs around 130 mg of pollen, which early flowers provide, to become a summer bee. As none of the early flowering plants is a significant source of nectar (and

hazel has no nectar at all), the bees will have to water down any stored honey as a nectar substitute. And if you have taken off all their stores, you will have to provide some thin sugar syrup in a feeder on top of the hive as a substitute for that.

Early nectar sources

It is only with the coming of the dandelion that the first major source of native nectar becomes available for the bees to collect and use as instant food or transform into honey for use on a rainy day. The dandelions will flower through to the very end of the year, when the last of the honey is capped and the bees have hunkered down for the winter. If you have dandelions on your lawn, try to leave them for the bees; they can be a year-round staple.

The season always starts earlier in the city. The climate is modified by the warmth of the buildings, and the multi-cultures of gardens, containers, and window boxes means there are always early bloomers. As spring arrives, I often find the most unusual colours being brought into the hive from exotics found by adventurous foragers. The vivid blue of Siberian squill may not be uncommon in the United States, but it was a surprise to me to see it coming into a hive one sunny March day in the middle of London.

Soon the fruit trees will be in bloom,

*Hardy geraniums
are a rich source of
nectar throughout
the summer*

providing both pollen and nectar, and
you will know the season is in full swing.
Apples are a terrific source of both high-
protein pollen and sugary carbohydrate-
rich nectar, and crab apples a perfect
urban tree, easy to grow, requiring little
if any attention, and providing lots of
flowers dripping with nectar and rich
with light olive coloured pollen. Cherries
are especially in need of bees for cross-
pollination, with trees of different sexes
relying on the bees to go between them
if they are to produce cherries. Cherry
pollen is brown, the pollen of plum trees
is grey, of apples a yellowy white, and of
pears a reddish yellow.

Tough urban plants

Lilac is another early flowerer and one
I often plant, along with lavender and
rosemary, close by my urban hives.
Buddleia is a ubiquitous London plant,
self-seeding and finding a foothold in
every nook and cranny, on rooftops, and
along train tracks. It is commonly known
as the butterfly bush, but bees love it
and as I wander round the city or glance

out from a train carriage or the top of a
bus, I will keep an eye out for scout bees
coming to the plants, checking out what
is available, and returning to the hives
with news. Foraging bees will appear a
short time later and work the flowers.
Campanula with its blue flowers is a
virtual weed on the streets of London,
and erigeron self-seeds in the cracks
between paving stones and is the easiest
of plants to grow; its daisy-like flowers
can be seen all summer long.

Bees see a different colour spectrum
to humans, shifted up a colour. While
our rainbow of visible colours goes from
red, through orange, yellow, green, blue,
and indigo to violet, bees start at orange
and go beyond violet to ultra-violet. A bee
cannot see red any more than we can see
ultra-violet. There seems to be something

Foraging on a rose

about the top end of this spectrum of colours that they are particularly drawn to, hence the ultra-attraction of lilacs, rosemarys, and lavenders. All these are perfect urban plants, tolerant of neglect, thriving in pots, enjoying the exposure of rooftops and balconies.

Urban trees

By May the 'flow', as beekeepers call the availability of nectar, is at its strongest and the bees at their busiest. The bigger urban trees are now in bloom. Horse chestnuts (*Aesculus hippocastanum*) and limes (*Tilia* spp.) are two of the most important sources of forage in the middle of London—the great trees in all the parks, royal and common, many of them having been there for decades, mature and towering in the sky. Horse chestnuts can be 75 feet (23 m) tall, with an equal spread; limes, long a park favourite, are amongst the tallest trees in Britain, reaching a height of 148 feet (45 m). These huge trees dominate much of London. A reason we see so few bees in London, even though there are many hives, is because they are far above us, foraging in the trees. In the Royal Parks alone there are 135,000 trees covering 1,100 acres (450 hectares); and of those, fifteen hundred are 'veterans', some hundreds of years old.

From plants to honey

It is only the older bees that leave the hive and forage for pollen and nectar. For the first half of their short lives, about three weeks, the bees work in the hive as house bees. For the second half of their lives, the other three weeks, they are out on the wing—some scouting, some protecting the hive, but most foraging. Once the spring weather is warm enough to wear a short-sleeved shirt, it is warm enough to open the hive and look inside. If you open up the hive and look into the brood box, you will see the pollen that has been taken off the foraging bees by

Horse chestnut

the house bees and packed into cells near to the brood, close at hand to feed the young. The horse chestnut pollen and nectar are both a rich dark brown, like the chestnuts themselves. The lime is perhaps the palest of forage, the nectar almost transparent and the pollen almost white. Some blossom is rich in pollen, some rich in nectar. A few have both. The nectars will have different colours and so will the honey made from them.

The stored pollen has a matt look to it, and the cells the look of a hexagonal paint colour chart. Unlike the pollen, the nectar glistens in the cells it is stored in, catching the light as you look into the frames. Some bees specialise in carrying pollen, others in bringing the nectar back into the hive; a few will bring back resin for use as propolis, the dark, sticky substance used to repair the hive and close up any gaps that might let in rain, wind, sunlight, or predators. The nectar-collecting bees ingest it and carry it home in a special honey stomach. As the bees ingest it, the nectar mixes with enzymes and starts becoming honey. A bee will carry almost half her body weight, as much as 40 mg, of nectar back to the

hive. There it is passed from forager to house bee, regurgitated into the honey stomachs of the bees back at the hive, and the process of making honey continues. The house bees put the nectar into wax cells for storage and reduce its water content by fanning it, the excess lost to evaporation. When the water content is reduced to 18 per cent, the process of transforming the nectar into honey is complete, and the cell can be sealed with a thin layer of wax to protect the honey until it is needed.

A honey bee will visit anywhere between a hundred and a thousand flowers on every flight away from the hive. And that same bee will make twenty flights a day. Much will depend on how much nectar there is available and how near to hand; nectar dries up quickly on hot days and when rainfall is scarce. Some flowers require more work than others: a bee may spend just sixty seconds on an apple blossom but twice that on a raspberry blossom. As you watch a bee foraging, you know that the longer she spends on a blossom, the more nectar there is to collect. It takes twice the time to collect a load of nectar

Campanula and erigeron

as it does a load of pollen. When the nectar is in full flow, a strong colony can fill a super with honey in just a few days, a reason to be sure there is enough room in the hive for their expansion. Better too many frames in the supers than too few.

Theoretically, the more flowers there are close by the hive, the less flying the bees will have to do. My observation, though, is that the bees will fly far from the hive to forage when they can and leave the blossoms close by for those days when the weather is inclement. They will fly up to three miles if they have to, but that is not very energy efficient—like driving miles to fill up your car and using all the fuel on the journey to and from the gas station.

Pollination corridors

Planting for bees, even if you are not a beekeeper yourself, is an important task. The more planting, the more bees can be sustained in urban environments. The more bees, the greater the gene pool, the more colonies for bees and drones to breed; and the more breeding, the stronger the bees. The queen from a solitary hive in the middle of town has little chance of finding mates. It is also important for the foraging bees to be able to move from plant to plant, hence the importance in towns and cities of 'pollination corridors' for the bees to fly along, foraging as they go.

In England, beekeepers talk of the 'June Gap' when there is little for the bees to forage on, and cotoneaster becomes an important food source. Urban beekeepers, with the rich variety of forage in the city and well-planted and -watered parks and gardens, suffer much less from the June Gap than rural beekeepers do. Thistle-like globe artichokes (*Cynara* spp.) are good to plant for this time of year along with mourning widow (*Geranium phaeum*) with its almost black flowers that keep bees busy from May to July.

Summer berries

As summer comes to its height the berries are in flower, raspberries first and blackberries a little later, both

From left to right
*Honey bees foraging on
sunflower, lavender, and thistle.*

with grey pollen. They can be in flower throughout the summer months, along with hawthorn and clover and foxglove— all important sources of both nectar and pollen. But by now the bee season is coming to a close, the numbers in the hive reducing and the bees concentrating on building up the last of their honey stores. For many English beekeepers July is the time to begin to harvest the honey, but just as the season starts earlier in the city so it finishes later. There are always some blossoms to be found on a sunny day late in the year, and the hive will take the opportunity to build up stores whenever it can.

As summer fades, the number of bees in the hive declines dramatically. Winter bees replace their summer sisters as they die off, and the colony prepares for the long, cold, dark months ahead. There will still be some sources of forage, even into the autumn and winter, ready to be collected by intrepid bees on a day sunny and warm enough. Sunflowers (*Helianthus annuus*) will be covered in bees from midsummer through to the autumn. The common honeysuckle (*Lonicera*

TOP TIPS

BEE-FRIENDLY GARDENS

- Every plant helps. Just one lavender in a pot is worth planting.

- Try to have a variety of plants to provide forage through the year.

- Free your garden of pesticides.

- Keep a pond or other water source for bees to drink from.

- Leave a wild area in your garden.

- Put up bee boxes to encourage solitary bees.

periclymenum) blooms as late as October. Ivy will be the last new source of forage, right up to winter time. The final dandelions will also be out as the hives close down and the bees begin to huddle together in a tight cluster against the cold.

By the end of the season a hive of bees will have collected 44 to 125 pounds (20 to 57 kg) of pollen, just a little of which will be stored for the winter months. They will have stored plenty of nectar, converted into honey. As the bees settle down for the autumn it is time for the beekeeper to do the foraging and harvest the honey.

SOME FAVOURITE BEE-FRIENDLY PLANTS

Spring
Crocus (*Crocus* spp.)
Winter honeysuckle (*Lonicera fragrantissima*)
Apple (*Malus domestica*)
Bluebell (*Hyacinthoides non-scripta*)
Willow (*Salix* spp.)
Raspberry (*Rubus idaeus*)

Summer
Giant viper's bugloss (*Echium pininana*)
Lacy phacelia (*Phacelia tanacetifolia*)
Poppy (*Papaver rhoeas*)
Lavender (*Lavandula angustifolia*)
Wild marjoram (*Origanum vulgare*)
Meadowsweet (*Filipendula ulmaria*)
Holly (*Ilex* spp.)

Autumn
Michaelmas daisy (*Aster* spp.)
Goldenrod (*Solidago* spp.)
Sunflower (*Helianthus annuus*)

Winter
Winter cherry (*Physalis alkekengi*)
Wintersweet (*Chimonanthus praecox*)
Christmas rose (*Helleborus niger*)
Ivy (*Hedera helix*)

And throughout the year
Dandelion (*Taraxacum officinale*)

Container planting for urban forage

"The complexity of flavours and scents in urban honey far exceeds that of honey from the countryside."

HONEY HARVEST

Perhaps I had underestimated my first honey harvest: I came prepared with only a small tin bucket. 'You'll never get it all in that!' John's voice boomed, and so saying, my mentor lugged a huge old plastic container that had once held Turkish pickled gherkins out of the back of his battered old estate car. Bits of hive, equipment, and even a few confused bees were always in the back of his car and today was no exception. 'It's going to be messy,' he had warned me, and messy it certainly was.

There is a little shed in the corner of the garden of the Natural History Museum. We heaved the supers full of honey up and off the hives and took them inside the shed. I had been down the day before to put 'bee escapes' into the crown boards and then place the crown boards between the supers and the brood boxes. Overnight any bees in the supers would have travelled down to the brood boxes through the valve-like escapes and been unable to return. So there were very few left in the honey-filled supers that we took off. And my, were they heavy! A super full of honey can weigh more than 35 pounds (16 kg) and lifting them off can

Honey from many different plants

Bees on a queen excluder

be back-breaking. Getting the supers off the hive can be hard enough, but the really difficult bit is getting the honey out of the supers.

Uncapping

The bees put a little wax cap on each cell when it is full of honey. To get the honey out you first need to take off the caps. Of course bees can uncap the honey themselves. When they need to feed on their stores they will neatly take off the wax cap, one cell at a time, and remove the honey. Through the winter months little piles of tiny wax flakes can be found at the bottom of the hives, an indicator (if you remove the tray from beneath the mesh floor) of whereabouts in the hive the colony is.

There are many ways for beekeepers to do the uncapping, and for my first harvest I used the simplest, an ordinary serrated kitchen knife. It does take some practice, and my first attempts saw broken comb, wax, and honey everywhere.

Eventually both sides of the frames were uncapped, and we were ready to spin. A honey spinner is like a dustbin with a wire container inside on a central pivot to hold the frames of honey. John had brought a spinner with him in the car, and we loaded up the first batch of frames.

Checking frame of capped honey

Our hand-powered spinner held just four frames, but there are other monsters that hold many more and are operated with an electric motor. First frames in the spinner, I turned the handle and the frames began to spin. Slowly I increased the speed and the centrifugal force made the honey fly out of the comb in fine threads of liquid gold, which stuck to the side of the extractor and slowly dribbled down to the bottom. The smell was quite wonderful. And dipping my finger into the comb, I could get my first taste of my first honey harvest.

It was hard work. Four frames at a time we extracted the honey, turning the handle, slowly at first but then building up speed, careful not to go too fast lest the combs break. The wire in the wax foundation helped keep them together, but even that is not enough to hold the wax in place if the spinner is turned too quickly. By the time we had finished there was honey everywhere. All over the table, on the floor, and in a bucket filled with the cappings we had removed.

But there was also an old Turkish pickled gherkin container filled with frothy honey that had come out through a tap at the bottom of the spinner. It was impossible for me to lift on my own. We had harvested nearly 100 pounds (45 kg) of the most natural food on the planet. All I had to do now was take it home, filter it through a sieve to take out any bits of wax and the odd dead bee, and bottle it. But what was it that I had in my giant bucket? What exactly is honey and how do the bees make it?

What is honey?

The nectar that the bees collect to make honey from is a dilute solution of the sugar sucrose. About 80 per cent of nectar is water and the rest sucrose. The enzymes that the bees add to the nectar convert that sucrose into other sugars—mostly fructose and glucose. While nectar is 80 per cent water, honey is 80 per cent sugar—just 1 per cent sucrose but around 38 per cent fructose and 31 per cent glucose, amongst others. It is the ratios of sugars, one to the others, that gives honey its particular sweetness. Some of that glucose is then converted by other enzymes into gluconic acid and

hydrogen peroxide. These give honey its low pH level. The hydrogen peroxide gives some antimicrobial and antiseptic properties to honey.

There are about 2 per cent of other things in the honey which add to its taste and nutritional properties—minerals, vitamins, pollen, protein, and amino acids. The bees have reduced the 80 per cent water content of the nectar to no more than 20 per cent water in their honey. More water than that, and the honey can ferment.

It is only the capped honey that you want to harvest. The bees will not cap the honey until the water content has been reduced to below that crucial 20 per cent. You can check your honey with a refractometer, which will give you its exact water content. Or you can trust the bees to do it for you and harvest only capped honey.

Colour and taste

Just as you can tell what bees are foraging on from the colour of the pollen they bring back into the hive, so too you can tell from the colours of the honey

Left
Bees capping honey

Right
Bee making honey

that they make. There may be a subtle rainbow of oranges, browns, yellows, and creams in the hive, different colours in different supers and even in different frames as the nectar brought in reflects the changing flora available. And as well as different colours there are different tastes and smells. The complexity of flavours and scents in urban honey far exceeds that of honey from the countryside.

Urban bees are better fed than their country cousins and as a result produce better honey. Where the world has dwindled towards monocultures, the honey bee has become restricted to a very limited diet. Vistas of oil-seed rape (canola) cover much of southern England, and there the bees must survive on the little yellow flowers that bloom in the spring. The monocultures of the United States change across the country, from the earliest almonds and citrus in California through to cranberries in New Jersey and Massachusetts later in the year. Although they all need pollinating, the flowering seasons are too short to sustain populations of honey bees. Instead the bees have to be moved from farm to farm and from state to state on giant trucks piled high with hundreds of hives. In China the hives travel at a more gentle speed on boats up the Yangtze river, following the climate, the season, and the crops.

Some honey farmers follow the blossom on a smaller scale, in Scotland taking their hives to the heather for the bees to produce the heather honey that will sell at a premium price. But while there might be markets for the particular tastes of monofloral honeys, or for manuka honey with its supposed special health benefits, the bees that make these honeys have very restricted diets.

The humble urban bee, though, lives life as a bee's life should be lived. Unlike other bees, which are especially adapted for particular plants, the honey bee is not a specialist. *Apis mellifera*, like the beekeeper, is an omnivore, adapted to feed on and pollinate a multitude of flora. For millennia that is what she has done and for centuries she has done that in the company of man. Only in towns and cities can a bee now hope to find a rich multiflora wherever she flies in a three-mile radius of her hive. That variety of nectar and pollen makes for healthy bees and the most complex and healthy of honeys.

Urban honeys

At London's Borough Market you can find honey of every colour and flavour harvested from the hives of the city. The darkest honeys are the chestnuts, with their rather bitter tastes, made by the bees early in the season. The lightest are the ivy honeys, almost white or translucent, made at the very end of the year. In the middle are blackberry, dandelion, lavender, and rosemary. But most urban honey is a mix of all these tastes and colours and many, many more, all the weeks of spring and summer and all the varieties of flora in the city brought together in a single jar. Some of it will be very runny and some will be thick, 'set' as the sugars in it crystallise, and these characteristics will also be dictated by the nectar from which the bees have made their honey.

Urban honey is very popular and the honey at Borough Market soon sells out. Some buy it for the health benefits, local honey seen as a protection against hay fever, eczema, and other allergies. Some buy it as part of a healthy life style and a connection with nature in the middle of town. Others make a purchase for the taste or the novelty. For the beekeeper selling honey there are one or two legal requirements. It has to be sold in jars of certain sizes, it has to have the name and address of the beekeeper on the jar, and it has to have a 'best before' date. But the honey in the jar is just as it has come from the hive, a perfectly natural product that will last for years, centuries, even millennia. There might have to be a 'best

A jar of honey with comb

before' date on the jar, but honey has been found in tombs of the pharaohs and proved to be perfectly edible.

Ways of harvesting honey

There are many ways of taking off the honey from the hive. I was lucky with my first harvest and it all went smoothly, if messily. Some beekeepers do it just a frame at a time, shaking off the bees and transferring the frames to a new box for taking away.

But if you have more than one or two hives it is often best to try to take off all the honey without removing the frames from the supers and with no bees in the boxes. The trick is to do it without the bees noticing, and for that you need to be sneaky. The bees are constantly moving to and fro in the hive. When they move down to the brood box you do not want them coming back up. Hence the 'bee escapes' that I used in my first year, and the various other devices that have been invented to ensure there is a one-way traffic of the bees when you want to clear the supers to take them off the hive and away for honey extraction. But bees being bees, they always seem to find a way around any human intervention, and in the end there will always be some bees who do not want to leave the honey and have to be shaken, blown, stroked, and flicked away, or even driven down into the brood box with scented sprays.

One summer at the museum I had the best harvest ever on one hive. Five supers heavy with honey. One by one I took them off and piled them to one side of the hive. I pushed little plastic bee escapes into the holes in the crown board, placed it on the brood box, and then replaced the supers full of honey and of bees, put back the roof, and went

FROM COMB TO TASTING

1

Uncapping

2

Spinning

4

Jars

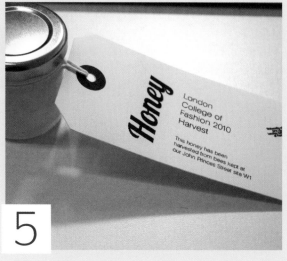

5

Label

3
Into jar

6
Tasting

away. Tomorrow, I thought, the bees will all be down in the brood box and I can come back and remove the honey. It had been a hot and sweaty job, the boxes of honey heavy and awkward to manoeuvre amongst the vegetation, and I had made one simple but costly mistake. I had not checked that the boxes were all put back exactly in place with no gaps between them. When I returned the following day the hive was obscured in a cloud of bees. They were everywhere. Word had got out (or rather smell had got out) that there was honey to be robbed from a small space between two of the supers. The bees were in a frenzy to take what honey they could back to their hives, including no doubt bees from that very hive taking honey out, down, around, and into the brood box by the front entrance. They were not keen on my getting near, even with smoker in hand.

I did eventually shift the top box back into position and close up the tiny gap that the bees were using to rob the honey. I let them calm down and drift away once they realised

there was no more robbing to be done. When all was quiet I lifted the boxes. There was no weight to them. The honey I had planned to take away and bottle had been taken away by bees themselves. A lesson learnt.

In a London teaching apiary, something similar happened around the same time. Supers were removed and placed carefully in the apiary shed for extraction. But without a secure cover. A week later all the honey was gone and just the dust and flakes of broken cappings remained as evidence that a scout bee or two had found a way into the shed and let the other bees know where their honey had gone. They had taken it all back.

I prefer to take off the honey at night—certainly late in the day once the bees have stopped flying. Otherwise they are inclined to follow you and interfere with what you are doing. With top bar hives, where the comb hangs down without a frame to support it, the bars have to be lifted out very carefully so honey-laden comb does

Supers from the hive

not break off. Once away from the hives, the comb can simply be mashed with a fork into a sieve or muslin cloth and the honey left to drip out over a day or two. You can do the same with framed comb if there is no foundation in it, or just cut out the honey on the comb and jar it like that.

The great advantage of the framed hive, though, is that you can give the frames back to the bees for the following year. Once my first harvest was over and the huge container of honey loaded up into the back of my mentor's car, I placed the supers and frames back on the hives. They were sticky with honey and a bit of a mess, but the bees, I was told, would clean them up. And that is exactly what they did. Coming back a few days later I found every super and every frame completely clean of honey. It was as if there had never been any. Every last drop had been taken down into the body of the hive. I lifted off the supers once more, this time seemingly as light as a feather, and took them away to store for the winter. Next season I would return them. All the comb that the bees had made in that first season would be ready for year two.

TOP TIPS

HARVESTING HONEY

- Prepare to get messy.

- Close the doors and windows.

- If you can, work after dusk.

- Check the regulations if you plan to sell your honey.

- Promote your location on your labels.

- Enjoy the aroma as you work.

"The months will be long and cold, and it may seem that you will never see your bees again."

WINTER

As the days begin to draw in and the colony reduces in size, it is time to prepare for winter. Eventually the bees will huddle down in a cluster to keep themselves, and their queen, warm during the winter months. The temperature inside the hive will drop to about 20°C and the cluster will move around the hive to feed on its stores of honey. If you have taken it all from them, then you will have to give them something as a replacement—sugar in the form of fondant, the thick white paste that is spread on cakes and buns. A local baker will often supply it, or you can purchase it from beekeeping suppliers in plastic bags. Just cut a hole in the bag and place it over the hole in the hive's crown board. The bees can come up and remove fondant from the bag and take it down to the cluster. Not nearly as good as honey, of course, but the bees will need it if there is not enough honey in the brood box to keep them going.

Pests and diseases

Now is the time to think about disease and pests, though an eye must be kept out for them

Hive in winter

throughout the year. The varroa mite is a tiny creature, the size of a pinhead, that invades a hive, multiplies in the brood, and lives on the thorax of the bee. The Asiatic honey bee evolved with this parasite and so is able to cope with the mite in its hive. But the varroa mite arrived in the homes of the European honey bee with devastating effect. Any beekeeper now has to be alert to keep varroa out of their hives.

Hives have been redesigned to help in this, with mesh floors through which the mite will drop if knocked off the bees. Unable to climb back into the hive, the mites can be collected on a tray under the mesh floor so that you can see how much of a problem you have. The little brown shiny creatures will glisten, like tiny pinhead-sized conkers, amongst all the other debris from the hive that has fallen through. There are various ways to kill them and to help the bees to remove them from the hive. There are chemical treatments, or you can dust the frames of bees with finely ground sugar and that will help dislodge them and make it easier for the bees to knock them off themselves.

There are diseases that the bee is prone to. Nosema is an illness whose dysentery-like symptoms can be seen outside the hives as the bees soil the hive and the landing board. There are horrible diseases of the brood, called European and American 'foul brood', though they know no national boundaries. These are so serious that in Britain you are obliged to notify your government bee inspector if you suspect them and, if necessary, they will destroy your colony.

With luck your hives will be healthy. Most are. Just as most beekeepers are. But once the honey harvest is off it is time to treat for any disease (so that no treatments end up in your honey) and ensure that your colonies are as strong as possible before the cold winter months.

Predators

It is also time to protect your hives from bigger predators. There are wasps, woodpeckers, foxes, mice, and badgers—all ready to eat your bees and your honey as highly nutritious sources of winter food.

I can hear a woodpecker at the Natural History Museum and occasionally see him. He does not yet seem to have noticed the hives, and I have got through all my seasons without him picking off my bees as they fly in and out of the hives, or worse, drilling a hole in the hive to get the juicy food out. Woodpeckers are a winter pest. In January and February when the ground is frozen hard, a beehive is an easy source of nutritious insects. Urban foxes can be much more of a problem throughout the year, quite fearlessly nosing into a hive, knocking over unstable ones. I came one morning to find a hive at Coram's Fields, a children's playground, with its super and roof pushed aside and the brood box exposed. The cold and wet had killed the colony. It might have been a human who had disturbed the hive (maybe the one who had stolen my scooter helmet a few weeks before while I was working on the hives), but more likely it was the fox that was standing on the wall next to me as I looked at the destroyed hive. I have strapped up all my hives ever since.

Straps will not stop a human of course, and may even tempt a teenager.

The hives at the Lillington Estate in Pimlico are protected against this particular pest. There are signs close to the hives warning of the bees and signs on the hives themselves saying DANGER BEWARE OF THE BEES in bold bright type. And the hives are strapped down within an inch of their lives. Despite all these precautions there are the occasional mornings when Jim the gardener arrives to find the straps removed. It is certainly not a fox who has done it. It must be a human, perhaps out of curiosity or maybe as a dare or a bet?

Smaller but no less of an urban pest is the house mouse, looking for a warm place to nest. They are as much a problem in a hive as in a house, and the best way to deal with them is to block up any hole they might get through. In the hive that means the entrance, which should be reduced to as small a space as possible during the winter, making it easier for the bees to defend and more difficult for the mice to squeeze through.

Worst of all are wasps. Like all bees except honey bees, colonies of wasps do not survive the winter. Once the queen has hibernated there is nothing for the

*South African hives protected
against honey badgers*

rest of the colony to do and they hang around in gangs, scavenging on anything they can find to eat. The sweeter the better. So they will invade your picnics and if they get a scent of honey they will invade your hives. Wasps can destroy a hive in a couple of days. Just as you are relaxing at the end of the season and decide to go away for the weekend, the wasps will arrive and clean out your hive. They will eat anything. Not just the honey but the bees as well. The bees will do their best to defend against the predators. A small hive entrance will help, as will wasp traps easily made out of old fizzy pop bottles. You can watch the bees literally wrestling with wasps around the hive, grappling each other on the ground. Give them whatever help you can.

Losing your bees

It is always sad to lose your bees. However careful you are with your husbandry and management, not all your colonies will survive. It has been the collapse of colonies that has led to the growth of interest in bees and beekeeping over recent years and brought many new beekeepers into the hobby. The causes of colony collapse are many and varied, but there are some that can be kept at bay. If you keep your bees fed over the winter, and protect them against disease, the chances are they will survive through to the spring.

After the regular weekly checks on the hives in the spring and summer, as the days get shorter and the temperature drops, reading the hives, strapped up and sealed against predators as they are, becomes much more difficult. But it is still possible to monitor the activity of the bees in the winter cluster. You can listen for them by pressing your ear on the side of the hive, you can feel the heat coming off the hive, and you can see the tiny fragments of wax that drop through the mesh floor of the hive as the colony uncaps stored honey. The pattern of wax pieces on the tray under the floor also gives you a sense of how large the colony is and where in the hive it is clustered. And you can heft for stores, carefully lifting the hive at one side to detect its weight and therefore how much honey the bees still have left to feed on. If you

are feeding with fondant, a quick glimpse into the roof of the hive will show you whether the bees have eaten up into the bag of sugar. In the depths of winter it does not do to delve too far into the hive—and there is little to be done for the colony except to fend off woodpeckers and mice, to keep it dry, and to ensure it has adequate stores.

The season has finished. Time to clear up and plan for next year. There is mead to be made from any leftover honey, lotions and potions, soaps and salves, to be concocted from wax. The months will be long and cold, and it may seem that you will never see your bees again. But then one day the pussy willow will be in bloom and a foraging bee will be out bringing in the first pollen of the year. The new season will have begun.

TOP TIPS

WINTER BEEKEEPING

- Clean your equipment and store it away.

- Give your bee suit a good wash with plenty of washing soda.

- Protect your hives against winter predators.

- Keep your bees fed.

- Go to meetings and talks at your local beekeepers' association.

- Plan for next season.

REPORTS
FROM
THE FIELD

There are as many different ways of keeping bees as there are beekeepers. My story is just one of the many. You will find twenty-three other stories in the pages that follow. Each is influenced by differences of geography, culture, climate, and experience. All are fascinating.

Opposite page
The green roof of Hearst Magazines, London

1

A Mile from Main Street

Bees in the backyard

North Adams, Massachusetts

Tony Pisano spent most of his working life in machine shops.
Now fifty-seven years old, he took up beekeeping after being laid
off in 2002. 'I decided I had enough of working for companies,' he
says, 'and became self-employed. I started making wind chimes and
wooden toys and crafts. I am a musician and now make my living
playing contra dance music, teaching accordion, doing handyman
work, and keeping bees.' As well as harvesting the honey from his
ten hives he also makes hand-poured beeswax candles, lip balm,
and hand salve. 'I get great satisfaction in figuring out ways to make
things myself when possible rather than buying.'

Tony keeps his bees in four locations, but his favourite? 'Still
right here in my backyard.' That backyard is in North Adams, the
smallest city (pop. around 13,000) in Massachusetts, less than a
mile from the city's Main Street. 'Our street is all old mill houses,
formerly side by side tenements, so the houses are a driveway's
width apart. My hives are at the end of my yard, which is about
seventy-five feet from the house.'

Tony used his machine shop skills to help him get started. 'A
friend of mine had been keeping bees for many years and asked
if he could put a hive in my yard. I was kind of afraid of bees, but
my wife uses a lot of honey so we agreed to give it a try. At first I
watched my friend from a distance, then one day he brought me
an extra veil, so I watched up close, standing still, with my hands
in my pockets. After a few visits, he asked if I'd like to pull one of

Tony Pisano with his hives

the frames out, and from that point on, I was hooked. The next year I ordered five packages of bees and have loved it ever since. The first year, I made everything for two complete hives except for the frames. I even made my own smoker from tin cans and stuff I had around the house, and ground an old lawnmower blade into a hooked hive tool.'

The climate in North Adams is pretty unpredictable and so is the forage. 'I've kept track of some blooms over the years,' says Tony. 'Dandelions usually bloom around the first week of May, and we often have frost by the first week of September. In the last two years, heavy frost has held off till the end of September, and last year dandelions were a full two weeks earlier than usual. In 2009, it rained all spring and summer and was a poor honey year; 2010 was very dry, and if it wasn't for rain and a good flow in the fall the harvest would have been a flop.'

Tony hiving his bees

Red maple is an importance source of forage for the bees. 'Linden is one of the main trees close by, and the weeds are abundant. We let our front yard grow pretty natural so tansy, milkweed, goldenrod, and ragweed thrive. I've planted some buckwheat in the past and we have vegetable and flower gardens, plus six fruit trees and eight blueberry bushes. The sweet clover grows wild around here, and there is a good stand of knotweed across the street on the riverbank. There is also sumac. Unlike some areas of the country, our flows seem to be more drawn out. There's usually something in bloom for the bees.'

Beekeeping has given Tony a new life and there is much he loves about it: 'I love the smells of a healthy hive, seeing the bees interact with each other, sitting by the hives in the early morning with a cup of coffee watching the guard bees check the incoming bees, seeing a bee chew its way out of a cell. I love the sound of the hive happily buzzing, hearing a piping queen for the first time, and the sticky joy of extracting honey. I love selling at the farmers' market and talking to the customers. I love going into the classrooms and talking to kids about honey bees. I love talking to the public and encouraging them to not use pesticides and plant more bee-friendly plants. I love the smell of my beeswax candles.'

The Lure of the Yellow Margin Orchid

Traditional Japanese beehives

Kyoto, Japan

2

Japan has a long history of beekeeping, dating back to at least the ninth century, when people came from all over Japan to present honey as an offering to the emperor. Hives developed in the twelfth century. While European beekeepers were using skeps and burning sulphur to kill all the bees in the straw hives so they could get to the honey, the Japanese had already invented both upright log hives and a multi-tiered hive of piled boxes, very similar to the Warré hive, and so were able to extract the honey without killing the bees. Like other Japanese beekeepers, Yuichi Shiga still uses these traditional hives in his apiaries in Kyoto. The bees build their comb down from box to box, the queen laying brood at the bottom and the honey being stored at the top. There are no frames, just boxes packed with wild comb.

When it comes to harvesting the honey, the bees are not smoked but tapped. After removing the roof, Yuichi lightly taps the lid of the hive with a short bamboo pipe to chase the bees deep into the bottom of the nest. He also blows on them through the pipe as added encouragement for them to go. Once the box of honey is empty of bees, a length of wire is used to slice through the comb between the bottom of the box to be removed and the one below it.

Yuichi Shiga's traditional Japanese beehive

The box of full comb is then lifted away. The roof is replaced and the bees are left in the hive after minimal disturbance.

There are plenty of wild honey bee colonies in Japan, mostly living in trees but also making their nests under floors, in ceilings, and even in graves. Yuichi began keeping bees as a way of rescuing wild colonies whose nests are often destroyed by other 'killer' insects as well as by humans. There is no need to buy packages or nucs of bees in Japan because the swarming of wild bees provides all the colonies needed. Yuichi houses swarms which he lures by planting *Cymbidium floribundum* (the yellow margin or golden leaf-edged orchid) next to empty hives, reckoning that during the season he has a fifty per cent chance of a swarm entering and making its home in one of these baited hives. When they are not feeding on orchids, these Japanese honey bees forage on Japanese chestnut (*Castanea crenata*), field mustard, and clover.

Yuichi has sixty traditional hives in Kyoto and four guiding principles as to where best to site them: they should not be too close to houses, should provide clear flight paths for the bees, must be shaded from the strong summer sun, and be sheltered from the winds.

A Villainous Place for Bees

Top bar and Warré hives in an English seaside town

Deal, Kent

3

One of Elvin's Warré hives

The small town of Deal in the county of Kent on England's south east coast is the prettiest of places. Its Georgian streets and ancient inns make a close-knit warren of human habitation right up against the sea shore. William Cobbett called it 'a most villainous place'. The wind that hits the town from the sea can be fiercely cold and there are few days for bees to venture out in the winter. It is from here that William Penn set off on 31 August 1682 in his ship The Welcome, on a journey that would see him founding Pennsylvania. It is not fanciful to imagine him taking a hive or two of bees with him on the boat. Captain Cook chose the shipwrights of Deal to make the low-hulled boats he needed for landing on the shallow beaches of the antipodes. Perhaps he too took bees from here, to New Zealand. Kent is known as 'the garden of England', and while the hops that are grown here to make English beer are pollinated by the wind, all the fruit in the many orchards have always been dependent on honey bee pollination. There is a long tradition of beekeeping in this corner of England.

In a shed at the bottom of his garden, just a step or two away from the sea, Elvin Roberts builds hives. A young bio-science graduate, he took up hive-making on leaving university, convinced that there were better ways to keep bees than the ubiquitous framed

Elvin Roberts in his workshop

hives. He makes the long top bar hive and is the only person in England making the Warré hive. Elvin's hives have windows in them with wooden shutters, so that the bees can be watched and checked without resorting to opening up the colonies, something that is of great value on Deal's very many cold, wet, and windy days. They seem to work well. A few streets away Liz Turner has one of his top bar hives in the garden of her large Victorian house.

A problem for coastal bees can be lack of forage. All foraging must be done away from the coast; not for them the three-mile radius of an inland bee: there is pitiful forage out at sea. The odd windmill on the landscape is a testament to the force of the wind that the bees have to contend with, and the increasing dominance of oil-seed rape (canola) fields on the landscape makes for a honey that quickly crystallises in the comb and is near impossible to extract. But Liz has plenty for the bees to forage on in her garden.

'We have apple, lilac, plum and bay trees, perennial flowers including lavender and rosemary, as well as a changing array of annual flowers and vegetables. The north-facing strip of garden at the side of the house has a hawthorn hedge, cherry and crab apple trees, and shrubs which act as a very necessary windbreak—the wind comes from the north east across the sea straight from the North Pole. This description sounds rural but we are surrounded by bungalows and council houses and are ten minutes' walk from the centre of this small coastal town. Our beehive is perched up on a bank in the less sunny part of the garden, well sheltered by hedge,

Top bar hive in Liz's garden

with the entrance facing south west. The bank was there when we moved in. Our first thought was to level it, but it turned out to conceal huge quantities of reinforced concrete that were the remains of a wartime air raid shelter so it remains as a feature!'

It is a good location with benefits both to bees, beekeeper, and the rest of the family who use the garden. 'The bank gets enough sun to encourage the bees out to forage and is separated from the main part of the garden so the comings and goings of the human occupants of the garden don't cross the bees' flight paths. I can be totally opportunistic about any interactions with them, added to which, I find it very companionable to have them in the garden.'

Liz has packed her garden with suitable plants for her bee companions, but though she does sometimes see them visit the thyme, oregano, bean flowers, and rosemary that she has planted, they tend to fly up over the buddleia in front of their hive and away out of the garden, 'to who knows where. We are surrounded by homes with gardens and judging by the range of pollen colours I see on the bees' legs when they return, the pickings out there are good and varied.'

'Hospitality rather than harvesting' is Liz's beekeeping ethos, and she is not a great connoisseur of honeys. She did, however, take one frame of comb in her first autumn, leaving the bees with ten, and it yielded three small pots of clear, pale gold honey: 'One for my mother-in-law, one for my grand-daughter, and one for me.'

4 | Public Hives in a Public Park

Delaware Center for Horticulture

Wilmington, Delaware

Peter Lindtner helping winterise the hives

There are twenty-five city parks in the tiny U.S. state of Delaware.
Gardener Jacque Williamson is responsible for installing
sustainable, ornamental, perennial beds of flowers that, as well as
adding colour to the parks, are also pollinator-friendly for the city's
bees. Jacque has two hives at the Delaware Center for Horticulture,
in downtown Wilmington. 'Our workplace is adjacent to Wilmington
State Parks,' says Jacque. 'These incredibly productive hives, being
placed not only in the centre of the garden and next to a state park,
but also in the middle of an urban area in which there is always
something in bloom, produced a hundred and fifty pounds of honey
our first season.

'We chose to site the hives, on a south-facing hillside, for
appropriate sun exposure and drainage, but also to be just out of
the path of public visitors to our gardens. Our building and garden
lot combined make up about an acre of land, and our gardens
here are designed to be display gardens of hardy, urban, heat- and

Jacque Williamson's honey harvest

drought-tolerant ornamentals. The intention is to show that even in an urban landscape, you can have a beautiful, sustainable garden.'

The bees have plenty to forage on in these pollinator-friendly gardens. 'They particularly like the black locust trees (*Robinia pseudoacacia*) growing in the woods that are adjacent to us,' says Jacque. 'We have many species of rudbeckia, echinacea, asclepias, as well as a newly installed edible ornamental garden, which has lots of peppers, strawberries, eggplant, peas, and fruit trees.'

Twenty-six-year-old Jacque has a mentor, Peter Lindtner, who has been keeping bees for over sixty years, since he was a young boy living in Czechoslovakia. Jacque and Peter harvest over seventy-five pounds of honey from each of their two hives, which Peter reckons to be double that of any of the other hives he looks after—he attributes their great harvest to the 'urban effect', that something is *always* blooming nearby in the city for the bees to eat. Jacque has a further thought: 'The fact that they are placed smack-dab right in the middle of our own urban gardens helps, too.'

5 | Chicago Suburbs

Chris Albert and his self-built hives

Chicago, Illinois

Chris's hives in the snow

'**Beekeeping is an art,**' says Chris Albert, 'it takes time to learn to read the hive.' A beginner beekeeper, Chris and his family live and keep bees in a suburb of Chicago, with hives in their backyard. He describes his neighbourhood as 'an established community with lots that are fifty feet by one hundred thirty feet, so they are comfortable but by no means large.' They are certainly large enough to provide homes for hives without giving problems to the neighbours. Chris is a hobbyist woodworker, so when a friend with bees encouraged him to become a beekeeper he built his hives and all their components himself. Perhaps that is why they are so stylish, in pastel green with cartoon bees painted around them.

Chris sits on his front porch on balmy afternoons to watch his bees. 'On a warm summer evening, bees do it as well. On a sunny day the hives get hot inside so when all the bees gather at home in the evening, the house bees send the foragers outside so they do not overheat the hive. I have created a screen inner cover for the hives

Aleena stands by as the swarmy bees walk into their new home

and prop the crown boards open to let some of the heat dissipate on hot days.' While the summers can be hot in Chicago, the winters can be very cold and Chris wraps his hives with insulating wrap to help the bees keep warm. He also has to make an exit hole for the bees high enough up the hive not to get covered in heavy snowfall.

The Alberts' first bees were bought as packages, but the family have since successfully captured and hived a swarm. The swarm was caught in a bucket and taken to an empty hive. 'We placed a board on the front of the hive, then we dumped the bucket of bees on the board. The bees then send out scouts to find a new home. Since the hive was right there, a scout was bound to find it.' Within minutes the bees were trooping into their new home.

The bees forage in the backyards of this Chicago neighbourhood and the result is a rich 'spicy' honey, which varies in taste and colour both in different parts of the hives and at different times of the year.

Chris is still learning about his bees and his young daughter Aleena is learning with him. Their most difficult task? 'It is really hard to find the queen with so many bees in the hive, so you really need to learn to read the signs of her existence.'

6

Bee Paradise in the City of Angels

Rescuing feral colonies and finding them homes

Los Angeles, California

Los Angeles is bee paradise according to Erik Knutzen. Erik rescues feral colonies that have made their home in places they aren't welcome and transfers them to hives in his backyard. He finds his bees in some unusual places—one hive was rescued from a vacuum cleaner, one from a bird box, another from a kitchen vent.

Erik is in his mid-forties and a writer. With his wife, Kelly Coyne, he wrote the book *The Urban Homestead*. He uses Langstroth hives but, he says, with some critical differences. 'We don't use foundation, queen excluders, or treatments of any kind. We use only medium-size supers. In place of the foundation there is a small "starter strip" that gives the bees a clue as to what direction to build their own comb in.'

The hives are in Erik's small backyard surrounded by a six-foot fence on three sides that forces the bees' flight paths upwards. There is a brick patio next to the hives. In the summertime he enjoys setting up a chair next to them to watch the bees take off and land: 'Kind of like plane spotting!'

What do they feed on? 'It's hard to say. There's a lot of flowering plants and trees here of every variety. Right now [February] avocado trees are blooming as are stone fruit trees and some succulents and cacti. As an urban beekeeper, the funny thing is that there are probably far fewer pesticides used than in agricultural areas, so our bees are getting cleaner nectar and pollen.

'I like what Charles Martin Simon says in his articles about

"Beekeeping Backwards"—"it's not about the honey, it's not about the money." I think what he means is that if you focus on the honey, you won't be successful. Honey is a reward for good stewardship, not an end in itself. That being said, we will harvest honey if and when our hives produce a surplus that the bees can spare. I have not gotten honey out of my hives yet. There's a lot of citrus trees in the area, so the honey of other Los Angeles beekeepers I've tasted often has a citrusy tang. But the bees will also go after many other flowers, so you'll also get very dark honeys as well.

'I've really enjoyed being one step ahead of the exterminator and being able to rescue bees out of buildings and either keep them myself or pass them on to someone else who wants them. When our beekeeping mentor Kirk Anderson opened up the vacuum cleaner that contained our first hive, the sight of those bees clustered on the comb was like peering into one of the great mysteries of the universe.'

He has very few problems. 'We practice "Backwards Beekeeping". There's not really much to do. If a hive fails for any reason I can go out and rescue one of the hundreds of feral colonies in Los Angeles. Since we're keeping only the stronger hives, nature pretty much takes care of herself.'

7

Herbal Teas and Honey

A hive on an office balcony

Melbourne, Australia

The staff of Village Well were at first highly apprehensive when
Amadis Lacheta suggested keeping bees outside their office in Little
Bourke Street, Melbourne. Local beekeeper Martin O'Callaghan was
able to diminish their fears with anecdotes of his urban beekeeping
experience and found the right spot on the company's balcony.
With little through traffic from staff or visitors and protected from
adverse weather conditions, the balcony provides a safe haven
for the bees and no danger to humans. Village Well is a company
specialising in 'placemaking' which it describes as 'the art of
creating great places'. So siting a hive at their office sits comfortably
with the organisation's ethos.

The hive was made by Martin, who specialises in simple,
elegant top bar hives. The bottom of the hive is made from flywire
to provide ventilation and viewing access to the bees' activities
without needing to disturb the colony. The wooden hive and bars
are made from sustainably harvested timber and weather-proofed
with natural oils and wax. Amadis says, 'The scent of sweet nectar
from the hive that wafts into the office on warm afternoon air is
indescribably beautiful.'

Extracting honey from the top bar hive could not be easier. The
comb is gently removed from the hive using smoke to calm the bees,
broken into pieces, and placed in a protected sieve over a container
in the sun that allows the honey to slowly drip through.

Village Well's balcony

The hive provides a point of interest for the office, and besides enjoying it themselves, Amadis says that she and her colleagues 'look forward to sharing some fresh honey [and] balcony-grown herb teas' with their many visitors.

8

Bees at the Opera

Four Seasons Centre hosts rooftop hives

Toronto, Canada

When beekeeper Fred Davis approached the Canadian Opera Company about placing hives on the roof of their opera house, they were most intrigued. Fred explained what a perfect spot it was for urban bees, being so close to the flowering plants at City Hall, Osgoode Hall, and University Avenue.

The hives stand on pallets on the shingle-covered roof, a small bucket of water placed nearby with a mound of little rocks in it so that the bees can perch there to drink. There is a lot of wind on the roof and with little in the way of natural windbreaks to protect the hives, Fred has built a barrier around them to offer some protection from the biting winter winds, which can see temperatures drop to minus 22°C.

The downtown site at the Four Seasons Centre is close to work for Fred, who is a management consultant with a big city firm, so it is easy for him to drop in during the working day. Fred enjoys the escapism of his beekeeping and knowing that he is doing his bit to help the honey bee and his local environment. He had another motive when he started his opera house beekeeping. 'The Ontario Bee Act prohibits raising bees in certain areas and sets difficult guidelines for the urban beekeeper to maintain or comply with, so I wanted those who cared to know that bees can do well in a city.'

The hives are visible through the glass doors and the windows, which look out from the public spaces onto the opera's rooftop, so

that opera-goers can watch the bees as they sip their interval drinks. The opera company hopes that their honey bees will help diversify the types of species of pollinators and plants that currently live in the area, and feel it is another way for COC to be a—literally—vital part of the wider community.

9 | *Cocktails and Canapés*

Hives on green roof provide honey for hotel

Vancouver, Canada

Artist-painted hives at the Fairmont Waterfront

If you check into the Fairmont Waterfront Hotel in Vancouver you can refresh yourself on the deck of the swimming pool with one of their honey cocktails while you look out on the beehives just a few yards away. The Fairmont's bees are kept in six hives in the hotel's 2,100-square-foot rooftop herb garden. Built in 1991, it was one of Vancouver's first green roofs. The garden is home to over sixty varieties of herbs, vegetables, fruits, and edible blossoms; ten different species of local birds; and the bees. The sunken garden is three floors above ground level and faces towards the sunrise and the open waters of Burrard Inlet. Guests can watch the bees from the deck and even from the pool. Bees close to a pool can be a real problem as the bees drown trying to drink from it as well as disturbing the swimmers, but the Fairmont has found a clever solution: a bee font, a water fountain that lets a slow trickle of water run down stone faces, giving the bees

a source of clean easily accessible water so that they do not head for the pool. Guests, if they are not in the pool or drinking cocktails, can join the weekly hive inspections that are led by the hotel's director of housekeeping and 'Bee Guy', Graeme Evans.

Vancouver's climate is very mild with little to no winter snow and temperate summers. The temperatures, combined with plenty of rainfall, mean that flowers are in bloom from late March to late October, making for a long beekeeping season. Graeme says there are hundreds of plants for the bees forage on. 'But,' he goes on to say, 'the single largest crop we get a nectar flow from is blackberry—it grows wild everywhere and half of our honey comes from this source.' The bee garden is also a garden for herbs and fruit, so the bees can be seen foraging on mint, fennel, rose, lemon balm, dandelion, and the other plants around the hives as well as the blackberry.

The single harvest gives four to six hundred pounds of honey, which Graeme describes as 'mid amber in colour, and with a fragrant taste much like a fine wine, the flavour the result of a blend of nectars from over a hundred sources, but with points of blackberry, lavender, lemon, cinnamon, and mint.' Sounds like the makings for a great cocktail.

10 Neighbourhood Nursery

Family hives help family business

Seattle, Washington

It is the sound of the hives while she is working on them—'a humming vibration like the sound of the Earth's core'—that gives Vera Johnson special pleasure. 'It reminds me of a singing bowl . . . or what I imagine a foetus may hear inside the womb of the mother's blood circulating in her body. It is truly one of the most comforting sounds I've ever heard.'

Vera took up beekeeping because her family eats lots of honey— in baking, with cereals, and in teas and coffee, preferring it to sugar because of its health benefits. As the owner of a nursery and garden centre in Seattle, Washington, the northernmost major city in the United States, she needs the bees to pollinate all the plants that she sells. Village Green Perennials, West Seattle's 'Neighborhood Nursery', aims 'to help people use natural and safe garden practices, to grow and maintain healthy plants, and to bring birds and wildlife to their gardens'. As well as the antique roses that Vera specialises in, the list of bee-friendly plants supplied by the nursery is a long one, from blackberries, strawberries, and raspberries to apple, pear, and plum trees, squash, honeysuckle, borage, calendula, jasmine, hosta, salal, mahonia, and geraniums.

Keen to keep her bees as far from her customers as possible, Vera's hives are in the warm, sunny, north west corner of the yard between her house and her nursery. The hives are very close to her potting area so that she can watch the bees while she is working. Vera and her neighbours are part of a drive to promote local produce and services; the nursery is one of the many independent small businesses located

Vera Johnson with a swarm

all over West Seattle, and Vera sees her nursery and her honey bees as an important part of helping sustain the neighbourhood and the wider community. 'You can grab an espresso to go, pick up locally grown produce or freshly baked bread, shop for unique gifts or household items, or choose a good book for a day at the beach. You can shop for plants, get your hair cut, have your taxes prepared, or apply for a mortgage at your friendly community bank or credit union. By "Staying Local" you help build strong neighbourhoods.'

Part of staying local includes buying Vera's delicious amber-coloured honey, fragrant from the nursery's plentiful blackberries and roses.

11 Charlton Manor Primary School

Child-friendly beekeeping

Greenwich, England

Observation window in school beehive

Small children and bees might not seem an obvious combination, but at Charlton Manor Primary in Greenwich they get on very well. The bees live in hives in the playground of this inner city London school, located close to the O2 Arena between the Thames and the major A2 arterial road.

The headmaster, Tim Baker, took up beekeeping when a swarm came and attached itself to a wall next to the school's main entrance. 'There was panic from some staff and calls to close the school,' he recalls. 'The children seemed very interested though. When it was collected I found out that the bees were very unlikely to sting when they swarm. I realised how little I and people around me knew about bees even though we had always taught the children that they were important. I was concerned that the lesson the children had got from that close-hand observation of the swarm was that bees were something to be feared.' To dispel the message of fear, Tim set about

The school apiary

finding training for himself and interested staff so that they could set up a hive on the school grounds.

The school keeps the hive in its garden, facing a hedge and near a pond. The hedge sends the bees up and over the playground, and the pond provides a close water source. There are many flowering plants in neighbouring gardens, and there are parks nearby as well, so the bees are not short of forage. As an inner city school many of the pupils do not have access to gardens themselves, so the bees provide an important contact with nature for them.

The hives have windows on the side so that the children can see the bees at work without having to open them up. The headmaster is convinced that the bees are of great educational benefit: 'There are a number of children with behaviour issues in the school. They were given the chance to work with the bees. Their behaviour has greatly improved and they delivered a talk to the local bee club at its annual general meeting.' The school started with one hive, raised a queen, and now has two colonies. The honey harvested is bottled by the pupils and sold to raise money for the school.

At first parents asked the head if it was wise to bring bees into the school. 'I pointed out that we have a wonderful garden and there are already many bees in the school.' One of the pleasures for Tim as a headmaster has been watching children and staff overcome their fears. Another swarm arrived last year and everyone was calm. Two of the children helped collect it and put the bees in an empty hive.

12

The Bee Whisperer

Breeding queens in Georgetown

Washington, D.C.

It was the financial crash that started Jeff Miller off as a beekeeper. Jeff was in real estate, and his business, and the world, as he remembers it, 'turned to Armageddon' in 2008. He continues: 'With little real estate-focused work to do, and a few dollars saved I began to change my focus. I had always been a big foodie and an accomplished cook, so growing my own vegetables seemed a good transition. It was in gardening that I began to understand the importance of pollinators as my peppers bloomed and then produced nothing. So I bought a hive, found a supplier of bees, and jumped into this darkness. I have heard about the addictiveness of beekeeping, and I am exhibit one, as I caught the disease.' Jeff now has fifty hives all over the District of Columbia and nearby suburbs.

Jeff and his family live in the Georgetown area of Washington, D.C., a neighbourhood characterised architecturally by rows of attached houses of differing size. Their family house is seventeen feet wide and has a flat roof, accessible through a skylight, where they keep their hives. The height prevents neighbours from seeing the activity and raising the alarm. Jeff reckons it to be a perfect site for a number of reasons. Its hidden location prevents the bees travelling at pedestrian levels as they leave and return to the hive. Finally, there is an air conditioning fan coil unit that provides a steady stream of condensate water out of a pipe that he diverts to a bowl from which the bees drink.

Jeff uses frames without foundation in the brood area to promote natural cell size, and limits medication by using genetically strong

Jeff Miller toasting a successful harvest

bees and queens. His focus is in the breeding of bees and testing new queens. Honey, therefore, is a nuisance for this beekeeper, as it has to be extracted. He prefers to leave his honey supers in place unless the harvest is huge to give the bees food and protection from the cold of Washington winters, during which the temperature can hover around freezing for three months without letup.

Beginning beekeeping was not painless for Jeff: 'When I first got my bees as nucs, I was excited to get them into their new home. Being brave ahead of brainy, I went outside to move them in some shorts and a T-shirt, intending to pull each frame and transfer it to the hive. I used no smoke or veil. I thought I was the bee whisperer. I got stung at least thirty times by the end of it. My hand blew up like a couple of balloons, I got sick, and finally visited the doctor who prescribed me steroids. I am now more careful around the bees, although I still rarely wear a veil.'

13 A Native Garden

Karl Arcuri's desert bee hives

Austin, Texas

Day of the Dead beekeepers

It can get very hot in Austin, Texas, with 100°F days in July and August increasingly standard. The springs and falls, however, are fairly mild, and the winters rarely see snow or ice except for the odd freak storm that pushes its way down from the north. For Karl Arcuri the most difficult weather condition is drought: it is challenging to keep bees in the summer when plants are struggling to stay alive.

Karl had always been fascinated with insects, especially the social insects like ants and bees, and long wanted to have his own hive. But it was difficult to find anyone in Austin to help get him started. 'I read everything I could, but knowing the theory and putting it into practice without ever visiting a hive made me a little nervous,' he says. It was not until all the news about colony collapse disorder in 2009 that he decided to look again into keeping bees and found a beekeeping class run by Round Rock Honey Company, which gave him the hands-on experience to feel confident enough about starting his first hive.

Like many places in the United States, there is a local city ordinance

for urban beekeeping and Karl had to find somewhere well away from houses for his hive. It was a friend who came to his aid, offering his property about three miles from the state capital, a large piece of land that made it possible to place the hive well away from the house and nearby neighbours. For Karl, 'this area is quintessentially Austin, where folks are keeping a wide range of creatures from chickens to a pot-belly pig named Thomas who roams the neighbourhood for acorns.'

Karl has an interest in native gardening and knows exactly what his bees have been foraging on. 'I had a pollen analysis done for my first honey harvest, and it was predominantly crepe myrtle (no less than 75 per cent), with mesquite being second at 9 per cent. Austin is known for its spring wildflowers which start in March, so I'm hoping that my established hive will be able to take advantage of the abundant native spring blooms this year.'

The climate allows for two harvests a year. 'Last year I had about forty to fifty pounds of honey. The first batch was very light and floral from the crepe myrtle, but the second batch in the fall was darker in colour and had a richer flavour. I really enjoy the entire process of

Left
Big hive, little man

Below
Nuclear hive

opening and inspecting the hive. Each hive has its own personality that changes throughout the year, and it really makes you more aware of the changing seasons. Since I've started keeping bees, I notice what plants are in bloom and how long it has been since the last rain. The community aspect was an unexpected pleasure. All the neighbours are incredibly interested in the entire process, and most have come over at some point to see the hive. I've also joined a local Austin beekeeping group to help educate future beekeepers as well as swap information with experienced ones.' Karl and his wife are also known for throwing elaborate Halloween parties. Their favourite costumes see them dressing up as Zom-Bees.

Pollinating a Raspberry Patch

14

Retirement project for Ted and Valerie Rock

Chicago, Illinois

The raspberry patch in Ted Rock's backyard in the Chicago suburbs has flourished since Ted took up beekeeping. His bees are hived nearby and forage amongst the raspberry bushes. The bees have plenty to take back to their hives, and the pollination they do while at the blossoms results in heavy crops of berries. Ted's raspberry patch is a microcosm of the importance of bees to food production and the importance of planting to sustaining the bee population.

Ted works on the bees with his wife, Valerie; they seldom do anything to the hives without both of them being there. It is a perfect retirement project for the seventy-year-olds after a lifetime in the military and complements their volunteering work at the Greater Chicago Food Depository and Garfield Park Conservatory. The Rocks had always had an interest in bees, and it was a talk with the head beekeeper at the conservatory that got them started on their own hives. Concerned about decline in bee populations in the United States, they felt like doing their little bit to help.

The Garfield Park Conservatory, located ten minutes west of downtown Chicago, offers an array of urban beekeeping events and classes for all ages. Through the volunteer beekeeper apprentice programme, individuals can learn to maintain hives and harvest honey. Basic beekeeping classes are offered to the novice beekeeper, teaching them about beekeeping equipment, hive design and construction, bee biology and behaviour, bee management, and bee products. Additionally, an annual event for kids and families is hosted each

Meeting the bees at the Garfield Park Conservatory

July where attendees can 'Meet the Bees' up close and personal and discover how honey is made. The conservatory also hosts an annual beekeeping forum where bee experts come together for a day-long event that includes panel discussions about beekeeping in an urban environment. Garfield Park proved the perfect place for the Rocks to learn about beekeeping and now gives them an opportunity to pass their skills on to a younger generation.

Chicago has four distinct seasons. The winters start with cold surges in November and last until March, and even that month can see subzero temperatures for a day or two at a time. In the summer it will get and can stay above 100°F for even longer stretches. The bees cope well with these extremes, and with no farm crops anywhere about, survive on clover and Ted's raspberry bushes.

It was a very hard winter in Chicago at the end of Ted and Valerie's

first season, and many of the city's bees did not winter over. The Rocks lost their one colony. It was a six-hour drive to Long Lane Honey Bee Farms to pick up two new colonies the following spring. When they got home after the long trip, it was still too cold to install the packages of bees into the waiting hives so they spent a few days in the basement while the weather improved.

For Ted, the greatest pleasure is learning about the bees: 'I love to learn new things, and this is something I knew very little about. I find in conversation that most folks know nothing about beekeeping and it is fun to be the expert, especially since it is so important for people to know about the demise of the honey bee populations.'

15

Speke Hall

Heritage site joins campaign to save the bees

Liverpool, England

Close by Liverpool's John Lennon Airport is a timber-framed building that was a house long before honey bees were shipped to America. The present building dates back to 1490, and there were earlier buildings on the site before that. The airport may be just yards away, an industrial estate another neighbour, and the river Mersey the third boundary, but the gardens of Speke Hall provide an urban idyll for bees in the midst of all that surrounds them.

The gardens as they are now were created by Richard Watt between 1855 and 1865, and Tom Davies is one of a team of three gardeners who have undertaken substantial replanting in the spirit of Watt's mid-Victorian scheme.

In 2009 the National Trust in collaboration with local BBC radio stations established beehives in many of its properties to support its campaign to highlight the plight of the honey bee. Speke Hall was one of fifty National Trust sites to have hives for the first time. Gardener Tom Davies put two WBC hives in a fruit and vegetable garden area at the hall that the Trust was developing for community groups to use and for the public to visit.

Tom stood the hives on a square paving surrounded by chip bark from the Trust's own woodland management programme, built a wooden framework, and added netting to the sides to reduce wind turbulence around the hives. Tom chose the site because it is a sheltered area away from the main gardens and Speke Hall itself, and

The Speke Hall beehives

he felt that the fruit and vegetable garden seemed to be a fitting place to provide a home for the bees.

There are a lot of trees nearby for the bees to forage on, especially sycamore and lime; there are plenty of shrubs like mahonia, and in summer plenty of wildflowers. Despite being so close to the coast, the hives are sufficiently sheltered by both the netting and the many trees, and Tom reckons that 'bad weather blows over quite quickly on most days.' Having started with WBCs, Tom has moved on to Nationals which, like many beekeepers, he finds easier to manipulate, though he is providing them with gabled roofs so the two hive styles sit well together.

The honey has been a great success, says Tom. 'We entered competitions with it (coming third in the National Honey Show) and supplied our National Trust shop; the stock all sold out within a week! It has a clear, light, golden colour and seems to be well received by all who have tried a sample of it. It didn't have an over sweet taste and had hints of citrus to the palate. There's a real pleasure in tasting the honey and in the novelty of watching something you (and the bees) have produced go from the hive to the shop where it is sold to the public.'

Having the hives as part of a national project linked to the BBC brought an unexpected benefit for Tom: 'I have already achieved my fifteen minutes of fame by being on TV and radio and in the newspaper, all in the space of ten months!'

16

Victorian Suntrap

Bees in a basement garden

Kensington, London

The small garden of a basement flat, tucked away in the centre of London's fashionable Kensington, provides home to four hives and a haven for their beekeeper, Fiona Shannon. A bit of a suntrap, the garden is one of a long terrace of Victorian gardens, and the hives look perfectly suited to their setting, as if bees had been kept there since the house was built. But Fiona is a recent beekeeper. A graphic designer, she finds living in London quite hectic and for her beekeeping is 'the perfect foil to city life': 'It has a way of making central London feel like living in the country. Also you get to give something back. It's a very rewarding hobby. It's really relaxing watching the bees working. I love it when they are busy near the house working on the summer flowers and you think, "Those are my bees!" And there's a particular time of day (about 2.30pm) when all the new bees come out and hover about outside the hive to get their bearings, which is always fun to watch.'

Once they've got their bearings, her bees forage beyond the garden. 'I wish I knew where they went! There are masses of private gardens and a lovely garden centre, plus Hyde Park and Holland Park close by, so they are not short of forage. In my garden they love the hardy *Geranium phaeum* with its dramatic dark flowers, alliums, thistles, lavender, and rosemary, all the herbs. At the end of the summer they were all over the passion flower, which surprised me. They also love the ivy growing up the plane tree in winter. Right now [March] they are off finding crocuses and hellebores—they're all coming back with big balls of bright yellow, orange, and white pollen.'

Fiona's garden apiary

Her first two hives were standard Nationals but because London bees tend to be very prolific, Fiona soon regretted having made that choice of hives and moved to deeper brood boxes. She has yet to harvest honey, preferring to leave the bees their own stores over the winter.

Fiona has been quite discreet with the placing of her hives, keen not to upset neighbours living only a few feet away from the bees. 'I live in central London. My hives are at the bottom of the garden under a giant

Foraging in Hyde Park

plane tree and hidden from view. I love them being in the garden, as I can pop down and see them when I get home from work, and it's easy to spot if anything's amiss. I wanted them to be out of the way of the neighbours and sheltered and this is the perfect spot. You'd never know they were there until you get up close and hear them buzzing. They fly up and over the shed (into which I can retreat when they get very busy) and away across Kensington and Chelsea to forage. I have to be quite careful when I inspect so as not to alarm the neighbours! My main problem is avoiding barbecue time. I chose gentle colonies as we are quite close to other people, and apart from the immediate neighbours that I've told, others aren't aware of the bees at all. There was one time when I broke all the rules of beekeeping and ended up running down the garden with several angry bees inside my bee suit ... but that's another story.'

Bee House Sculptures

Artistic homes for native bees

Tucson, Arizona

17

Practical Art, a gallery on Central Street in downtown Phoenix,
Arizona, sells just that—art with a practical function. Seventy-five local
artists display everything from salad bowls to soaps and jewellery.
One of the gallery's most popular artists is Tucson-based Greg Corman
and his medium of choice is another man's trash: old rusty nails, car
parts, and other scrap metals are salvaged to adorn blocks of wood that
have been carved, burned, and burnished to become decorative works
of art. Sturdy enough to be displayed outdoors, Greg's sculptures are
also habitats to some of Arizona's thirteen hundred species of native
bees. As Greg says, 'These important pollinators do their work with
little notice and less credit while the honey bee, an imported species
from Europe and Africa, gets all the attention. While honey bees form
colonies as large as fifty thousand individuals, our native species are
mostly solitary docile creatures that nest in holes in the ground or in
dead trees and visit cacti, wildflowers, and native trees and shrubs for
pollen and nectar. In turn, they guarantee the successful fruiting of
those plants. They are unruffled by human presence and will sting only
if caught; and even then, the sting is much less potent than that of a
honey bee or wasp.'

Greg's sculptures from recycled wood and steel were first created
to support leafcutter bees. The wood is drilled with small tunnels that
the females use for nesting. 'Normally they would nest in dead tree
branches where beetles have drilled holes for them,' Greg explains, 'but
in urban areas we tend to remove dead material for aesthetic or safety

reasons and the bees go wanting for nest sites.'

A female bee will cut small pieces of leaves and shove them into the end of a tunnel and then deposit pollen, nectar, and an egg. She will repeat the process until there are several cells containing eggs in each tunnel. The bee then seals the tunnel with chewed leaves, resin, sand, or a combination of materials, and leaves the eggs to develop and hatch on their own in a matter of weeks or months. She might repeat this process in several tunnels during her life of one or two months.

Greg's design philosophy and practices can be summed up with the acronym **LEAF**:

Greg Corman's bee house sculptures

Local plants for beauty, toughness, and ease of care

Eco-friendly ways to save water, avoid chemicals, and nurture wildlife

Artwork to complement spaces indoors and out

Functional designs for comfortable living and low maintenance

'As an ecological landscape designer, wildlife habitat is a major focus of my work,' says Greg. 'Incorporating bee habitats is just one aspect. I also strive to make life easier for birds, butterflies, snakes and lizards, and many other desert creatures. All these efforts can be done artistically with plant arrangements, stacked rock, water features, and other garden elements. And that's where art meets ecology. Bee habitat sculptures are a great way to support pollinators that are essential for the health of our desert ecosystem. They look great simply as sculptures, too!'

18

Boston Blogger

Bees and a sustainable garden

Harvard, Massachusetts

Rochelle Greayer is a Boston-based writer and landscape designer.
Her daily gardening blog is dedicated to garden and landscape design.
She is a strong advocate of keeping bees as a vital part of sustainable
gardening.

Rochelle started and runs a farmers' market in her town of Harvard,
Massachusetts, and meets many beekeepers selling their honey on
market day. It is not only honey that is sold at the market but mead as
well, with mead-making workshops being a regular sellout. Rochelle
contracted with Fred Farmer of Nissitissit Apiaries to have two hives
in her garden for 'the season'. The hives are placed downhill from
but near to her large vegetable garden, and the bees are also in close
proximity to Rochelle's cultivated flowers and a meadow full of
wildflowers. It is difficult to imagine a richer and more varied forage for
a couple of bee hives.

Beekeeper Fred felt that the place would be sufficiently near to
the house but also unobtrusive, not interfering with Rochelle and her
family's use of their garden. For Rochelle, keeping bees would add a
new venture to her home life and would help the flowers to flourish
and the vegetables to thrive. Here's what she wrote in her blog, the May
day the bees arrived:

'Hive hostess' Rochelle Greayer

My phone rang at 5:15 this morning. Incredulous and startled, I ignored it, even though I recognised the name of the apiarist that sells honey at my farmers' market. I haven't spoken with him since last fall, when we agreed that he would install two hives in my garden this year.

Then, at 6:00 there was a knock on my door (thankfully, I was now up—but still in my PJs). I was greeted with a hearty 'Get up! It's the Honey Man!'

I am dealing with an utterly unapologetic Farmer.

And by 6:25 we became beekeepers—but only in the sense of 'we have bees on our property' and not in the sense of 'I know how to take care of bees' … so for now, I think maybe I should just call myself a 'hive hostess'.

It's 7:39 EST—how has your Monday morning been so far?

19

Town Bee and Country Bee

Jams, jellies, and honeys at Union Square Greenmarket

New York, New York

At Union Square Greenmarket in New York City, you can buy honey produced in the heart of Manhattan. Andrew Coté is a prominent fourth-generation beekeeper, with hives on rooftops across New York as well as in Connecticut, and the founder of two non-profit organisations, Bees Without Borders and the New York City Beekeepers Association. The association, which offers beekeeping classes and monthly workshops on topics ranging from swarm prevention to mead-making, has flourished since New York City lifted its ban on beekeeping. Through Bees Without Borders, Andrew teaches beekeeping as a method of poverty alleviation for under-served communities from the local (East New York, Brooklyn) to the far-flung (Iraq, Uganda, Nigeria, Haiti, and Ecuador, to name a few).

There are others selling local New York honey at Union Square Greenmarket. You can buy jars produced by Mary and David Graves from hives on the roof of a nearby hotel. You can also buy their out-of-town honey. Berkshire Berries®, owned and operated by Mary and David, is located in Becket, Massachusetts—a rural community noted for its lush wooded hillsides and cool summers. In 1978, when Mary and David produced their original three jams and jellies, they used native blueberries from the surrounding woods and raspberries and strawberries from their own backyard. They realised they needed to keep bees to pollinate their fruit bushes and were soon selling honey along with the jellies and jams.

*Berkshire Berries stall at Union
Square Greenmarket*

*Andrew Coté at Union
Square Greenmarket*

As well as selling the honey, jellies, and jams at Union Square Greenmarket, they have a family country store and gift shop in Becket. The two locations give them a special take on both urban and rural beekeeping. In the Berkshires they have to keep the hives away from black bears, in the city away from humans. The cold winters are the same in and out of town.

David Graves' father encouraged him to take an interest in nature, and beekeeping intrigued him as a child. It was not until he became an adult that he kept his first hive. He finds beekeeping a quiet, solitary relaxation. Colony collapse disorder has been a real problem for the Graveses, taking ten of their seventeen hives in one go. Honey production has dropped significantly in the last five years. David believes that cell phone towers cause bees to get lost and that climate change means no January thaws: the bees are unable to break the winter cluster and so starve to death.

The New York City bees forage on honey locust (*Gleditsia triacanthos*), sumac, linden, asters, ginkgo, Japanese knotweed, and whatever they can find on rooftop gardens. The various nectars impart very different tastes to the honeys on sale in Union Square; if you want a minty honey go for the Graveses' linden honey, but if you want something with the flavour of caramel candy, Japanese knotweed honey is the one to buy.

20 Baboons and Honey Badgers

Beekeeping in southern Africa

Port Elizabeth, South Africa

On the south coast of Africa from Cape Town to Durban the temperature barely falls below 20°C, even in winter. So the bees are out all year round, foraging according to the plants available. Port Elizabeth is a sprawling industrial city, spreading back from a seafront that is both a big international port and a holiday resort. Menno Alting gets called out all the time, especially in the height of summer (February and March) to collect swarms of bees. As in London, the calls are mostly to chimneys and building voids, but Menno must also board two or three boats a year. 'They just need a tiny hole to get in.'

Most of Menno's approximately five hundred hives are kept on his small holding on the city's edge. His hives can yield 44 pounds (20 kg) of honey each. Most of it is eucalyptus honey; the tall trees with their tiny flowers are all around him. Early settlers imported these 'exotic' trees from Australia, and many people now want them removed from Africa to make way for indigenous trees. But the bees love them. In the blazing sun, where you need protection from the heat more than from the bees when you are out working the hives, Menno stands under a couple of eucalyptus towering fifty or sixty feet above him. 'Listen,' he says, and you hear the steady buzz of thousands of bees working in the canopy above.

Menno has another job besides bees. He is a tree surgeon, cutting down unwanted trees in gardens across the city. For years the felled trees were wasted, but now Menno and his staff slice them into planks and make them into hives. It is a perfect recycling business.

Menno is not alone in his apiary pursuits. He and his fellow

beekeepers have found a way of managing the bees of a big industrial city without the sort of interference common in the UK and United States. Honey bees are indigenous to southern Africa. They are everywhere in town and countryside. There is no swarm control, but empty hives are dotted around the city for swarms to move into. 'I could put out a thousand hives, and in a week eight hundred and fifty would have bees in them,' he says.

The colonies are in balance, and there are many wild and feral colonies. Varroa seems not to be a problem. 'Yah, if the colonies are strong, they will deal with it.' A visiting bee inspector from England had been amazed to find just a few mites in hundreds of hives. Menno believes his success comes from minimal interference. He and the other South African beekeepers do not use queen excluders, calling them 'honey excluders' and regarding them as a constraint on the queen's ability to lay. A strong hive will lead to the queen laying in a few frames of the first super if she needs to. If the honey is in full flow the queen will retreat back into the brood box, and once hatched the super cells will be used for honey storage.

The only problems are drought and predators. Port Elizabeth lies between the sea and the desert of the Karoo. Minimal rainfall can make for minimal nectar. Predators come in the form of baboons and honey badgers, always in search of an easy meal.

Left
Hive on tyre stand

Above
Hives from recycled timber

21 Bees at the President's Palace

Hives in the heart of the German capital

Berlin, Germany

'My bees are for sure the best protected bees in Germany!' says Frank Hinrichs of his six hives in the grounds of the palace of the German president in Berlin. The Schloss Bellevue sits between the great Tiergarten park and the river Spree, which cuts through the city. The bees proved so popular that Frank was asked to put hives on the roof of another Berlin landmark, the city's cathedral, the Berliner Dom. Frank runs the nearby Citystay Hostel so can visit his hives on his lunch breaks or on his way home from work, fitting their management into his day job. All Frank's bees are in Dadant hives and all are Buckfast bees (originally bred by Brother Adam, a Benedictine monk in England's Buckfast Abbey), because, he says, 'I just felt that for me the Buckfast bees are more fun to work with.'

He also keeps bees in the garden of his home in the green suburb of Gatow on the border of Berlin and has hives at the city's beekeepers' association teaching apiary, in the beautiful, protected forest at Spandau in the west of the city. It is the city centre bees, foraging on the Berlin's famous lime trees, that give Frank the greatest pleasure, and provide a respite from the rest of the day: 'Working with my bees, especially at my city centre apiaries, is very relaxing; the sound of calm bees, the smell of honey and wax, the beautiful park around or the stunning view from the cathedral's roof—it's just unique to be with your bees in these locations.'

Frank Hinrichs and his bees, high above Berlin

22 | HK Honey Company

Beekeeping amongst concrete tower blocks

Hong Kong, China

Right
*Moving a hive in
Hong Kong*

Opposite
*Designer and beekeeper
Nelson Chan*

If ever a city was unpromising for the keeping of bees it is surely Hong Kong, with its seven million people crammed into concrete tower blocks on 400 square miles of the most densely populated land on the planet.

Yet it is here that product designer and beekeeper Michael Leung has established HK Honey, an organisation of beekeepers, artists, and designers that aims to communicate the value of bees and the benefits of locally produced honey. 'There are not really that many green spaces within the city,' says Michael, 'and so it was always going to be a challenge to see whether bees could actually survive and it would be possible to harvest honey in this city.' With its network of bee farms, a studio that designs products and services relating to urban beekeeping, and collaborations with various Hong Kong cafés, HK Honey has met that challenge. For Michael, it is about 'connecting local people with

local beekeepers and supplying them with local honey.'

As a group of artists and designers, as well as beekeepers, Michael and the other members of HK Honey make the most exquisite artefacts from the products of the hive, of which candles are the most popular. The group run regular candle-making workshops, each of which concludes with the eating of a special honey cake using their own honey as a natural and healthier sweetener. The recipe is from Bo, who works at Wontonmeen, a Hong Kong café; and all the other ingredients are also sourced from local suppliers, with milk from Kowloon Dairy and eggs from the nearby Chinese mainland (unlike most eggs eaten in Hong Kong, which are flown in from Thailand or the United States). The cake is served on special HK Honey trays that are made from the same wood that the group use to make their beautifully crafted beehives.

There is a meditative quality to Michael's urban beekeeping. 'When

Michael Leung leads a school tour of a Hong Kong urban bee farm

I'm working with the bees I move very slowly, I'm really relaxed, so everything's really calm, it's almost Zen-like.' He believes that the origins of food are now a mystery to most people and that city dwellers in particular are very detached from where their food comes from. It is important, then, that the urban bees of Hong Kong provide a link to another, rural, China. Michael recalls getting an email from a young woman. 'Her elderly mother was interested in keeping bees on her rooftop. Her mother grew up in the really rural part of China and her father was actually a beekeeper, so she has this natural connection to nature and also to bees. We moved a beehive to her roof, and it makes me happy to think that she can reminisce and think about her childhood, where she grew up on the farm.'

Bee Tourist Attraction

Glass-walled observation hive in botanical gardens

Turin, Italy

23

'Even the city of Turin can produce excellent honey,' says Lorenzo Domenis, who keeps his hives in the botanical garden at Turin's Parco del Valentino. The garden is between the river Po and the Corso Massimo d'Azeglio, one of the biggest and most important streets of this Italian city. 'We chose this area for many reasons, the primary of which is that it is in the city centre, and the surrounding zone has a rich floral impact. In addition (because one of our goals was the educational aspect), the park is visited by many tourists from April to September and beyond, and is close to most city schools, which facilitates the visits of students.' One of the hives has been especially made with glass walls so that when its covering is removed, tourists and students can see the bees at work.

Turin is Italy's fourth-largest city, in the north of the country close to the Alps. The geography makes for very cold winters and warm summers. The bees have plenty to feed on, from native limes and horse chestnuts to the many exotic plants in the gardens and greenhouses, such as alpine plants, medicinal plants, fruit trees, and tropical bromeliads, orchids, and aroids.

Lorenzo looks after the bees with his old friend Marco Cucco. They named their original five hives after their wives and daughters—Raffaella, Laura, Cecilia, Sibilla, and Elisabetta; as the hives have increased they have had to find names elsewhere, and these include Carlo Allioni, who founded the botanical garden in 1729.

Marco and Lorenzo harvest their honey three times a year, as

Lorenzo Domenis and Marco Cucco in Turin

Lorenzo explains: 'The first extraction in mid May is a light brown honey, collected at the end of the flowering of the horse chestnuts; the second is a clear yellow honey, collected in early June and characterised by the presence of *Robinia pseudoacacia* pollens; the third is a dark brown honey produced by bees in June and July, featuring a mix of mainly lime tree, honey locust, and chestnut. The first honey is very distinctive because its production is possible only in the cities, where streets and parks have a great density of horse chestnuts, very few of which are to be found in the countryside and villages of Italy.'

This region of Italy has a long history of beekeeping dating back to mediaeval times, which Lorenzo and Marco are keen to revive. They also have another ambition, to convert Turin into a 'self-sufficient eco-municipality', where apiaries sit alongside other initiatives such

The botanical garden's hives

as urban gardens, bike-sharing kiosks, and multi-storey farms. In a world that talks more and more about 'zero food miles' (pushing the population to consume food produced close to home), these two Italian city beekeepers believe they have proved it possible for even the most urban consumer to benefit from 'zero honey miles', and consume only honey produced in the very heart of the metropolis.

The two beekeepers have not only named their hives, they have even given a name to their honey. In local dialect *'L Buss Doss* means 'the sweet hive'.

RESOURCES

Australia

Beekeeping in Western
Australia
beekeepingwestaus.asn.au

Rooftop Honey
rooftophoney.com.au

Victorian Apiarists'
Association
vicbeekeepers.com.au

Canada

Canadian Honey Council
honeycouncil.ca

Ontario Beekeepers
Association
ontariobee.com

France

Association Terre
d'Abeilles
sauvonslesabeilles.com

Institut de l'abeille
cnda.asso.fr

La Société Centrale
d'Apiculture
la-sca.net

Germany

Bee Hotels
bienenhotel.de

German Beekeepers
Association
imkerverein-brb.de

Ireland

Federation of Irish
Beekeepers' Associations
irishbeekeeping.ie

South Africa

African Beekeeping
Resource Centre
apiconsult.com

South African
Beekeeping
beekeeping.co.za

Southerns Beekeeping
Association
beekeepers.co.za

United Kingdom

BeeBase
Government's National
Bee Unit website
secure.fera.defra.gov.uk/
beebase

British Beekeepers
Association
bbka.org.uk

Cymdeithas Gwenynwyr
Cymru / Welsh
Beekeepers' Association
wbka.com

Scottish Beekeepers
Association
scottishbeekeepers.org.uk

Ulster Beekeepers
Association
ubka.org

United States

American Beekeeping
Federation
abfnet.org

Beesource
beesource.com

Eastern Apicultural
Society of
North America
easternapiculture.org

National Honey Board
honey.com

New York City
Beekeepers Association
nyc-bees.org

International

Apimondia—The
International Federation
of Beekeepers'
Associations
apimondia.com

International Bee
Research Association
ibra.org.uk

Bees for Development
planbee.org.uk

Top bar hives

The Barefoot Beekeeper
biobees.com

Warré hives

Major Hives
majorbeehives.com

Journals and magazines

American Bee Journal
americanbeejournal.com

The Australasian
Beekeeper
theabk.com.au

Bee Craft
bee-craft.com

Bee Culture
beeculture.com

Beekeepers Quarterly
beedata.com

INDEX

Page numbers in **bold
italic** type refer to
pictures.

A

Albert, Chris 132–3
allergic reactions 41
Alting, Menno 166–7
apple 88, 91, 96
Arcuri, Karl 148–50,
  149
Australia 136–7
autumn 96

B

badgers 114
Baker, Tim 144–5
balconies 8, **8**, 12
bee box 95
bee escape 107
bee space 23–4, 31, 73
bee suit 40, 56–7, **58**
beekeeping season
  85–6
blackberry 92, 106
bluebell 96
bole **21**, 23
boots 40, 56, 57, **58**
brood 35, 55, 66
brood box **29**, 31, 72
brood frame **29**, 31
buddleia 88

bumble bee 35
buying bees 35–51
  nucleus 36, 38–9, 78
  packaged bees 44,
  46, **47**, 49–50, **149**

C

campanula 88, **93**
Canada 138–41
candles 26
Carniolan honey bee
  50
Carr, William
  Broughton 25
cats 17
cave painting **20**, 21,
  23
cedar hives 27
cells 72
  capped 66, 91, **103**,
  104
  honey 101–2
  manufacture 72–3
  nectar 91
  pollen 91
  queen cups 79
  uncapping 101–2,
  108
Chan, Nelson **171**
Chapple, John **14–15**,
  15–16, 26–7, 38–40,
  99

Charlton Manor
  Primary School
  144–5
cherry 88
chestnut 106
China 105, 170–2
Christmas rose 96
climate 14
clothing, protective
  40, 56–7, **57**, **58**
clover 94
coastal sites 128
colony 36
  creating new
  colonies 36–8
  expansion 69–72
  size 55
  splitting 38, 78–9
colony collapse
  116–17
comb 28, 72–4
  harvesting honey
  110–11
  uncapping 101–2,
  108
  wild 72–4, **78**
communication 19,
  40
Corman, Greg 159–61
Coté, Andrew 164–5,
  165
cotoneaster 92
crocus 86, 96

crown board **29**, 32,
  59, 66
Cucco, Marco 173–5,
  174

D

dandelion **86**, 87, 95,
  96, 106, 123
Davies, Tom 154–5
Davis, Fred 138–9
Deal, Kent 127–9
death 70, 71
diseases 38, 56,
  113–14
dogs 17
Domenis, Lorenzo
  173–5, **174**
drones 36, 81, 83
  life cycle 75
dummy board 65, **66**

E

eczema 106
eggs 31, 32, 53, 55,
  66, 70
  gestation 70, 75
Einstein, Albert 17
entrance 31
entrance reducer **29**
erigeron 88, **93**
Evans, Graeme 140

F

Farmer, Fred 162
filtering 102
foul brood 114
foxes 114, 115
foxglove 94
frame grip 65, **66**
frames 24, 26, 31,
   72–4, **78**, 111
   brood **29**, 31
   foundation 73, 134
   harvesting honey
      110–11
   hives 42
   honey spinner
      101–2, **108**
   inspecting 61, 65–7,
      66
   nucleus boxes 42
fructose 102
fruit trees 88, 127

G

geranium **88**, 92
Germany 168–9
globe artichoke 92,
   95
gloves 40, 56–7
glucose 102
goldenrod 96
gorse 86
government
   inspectors 114

Graves, Mary and
   David 164–5
Greayer, Rochelle
   162–3, **163**
guard bees 39, 66, 83,
   89

H

hawthorn 94
hay fever 106
hazel 87
heather 25, 105
Hinrichs, Frank
   168–9, **169**
historical
   background 21–4
hive tool 42
hives 19–33
   Beehaus **6**, **25**, 33
   box frame 17
   choosing 26–33
   Dadant 25
   development of
      modern 23–4
   flat-packed 28
   floor **29**, 31, 114
   inspecting 55–67
   Japanese 25
   Langstroth 23–5, **24**
   moving 48–51
   National **24**, 25, 27,
      30
   natural 26

People's 25
plan **29**
plastic/polystyrene
   6, **25**, 26, 27
siting 17, 28
skep **21**, 23
Smith 25
stands 28, **29**
straw **21**, **22**, 23
temperature 55, 72,
   113, 132–3
top bar 17, **25**, 26,
   32–3, **37**, 110, 128,
   129
Warré **25**, 33, **127**,
   128, **128**
WBC **24**, 25, 26
wind, protecting
   against **30**
windows **144**, 145
wooden 26, 27–8, **27**
hiving **15**, 40, 42, **43**
holly 96
honey 23, 35–6, 53,
   55, 66, 72, 86
   antimicrobial
   and antiseptic
   properties 103
   colour 103, 105, 106
   fermentation 103
   filtering 102
   harvesting 99–111
   health benefits 106
   keeping 106–7

labelling for sale
   106, **108**
manufacture 91–2,
   102–3, **105**
pH 103
runny 106
sale 106
set 106
sugars 102
taste 103, 105
water content 103
honey bee 35–6, **36**
   subspecies and
   cross-breeds 38
honey smell 105
honeysuckle 94–5, 96
horse chestnut 89,
   90, 91
horses 17
house bees 89, 91

I

inspecting hives
   55–67
Italy 173–5
ivy 95, 96, 106

J

Japan 25, 125–6
Johnson, Vera 142–3,
   143
June Gap 92

K

Knutzen, Erik 134–5

L

Lacheta, Amadis 136–7
landing board 28, 42, 71
Langstroth, Rev. Lorenzo 23–4, 44
larvae 55, 66, 70, 75
    feeding 70
    queens 79
lavender 88, 89, **94**, 96, 106
Leung, Michael 170–2, **172**
life cycle of bees 70–1, 75, 81, 83
lilac 88, 89
lime tree 89, 91
Lindtner, Peter **130**, 131
Liverpool 154–5
London 8, 12, **12**, **13**, **14**, 144–5, 156–8

M

manipulation cloth 65
manuka honey 105

marjoram 96
mating 81
meadowsweet 96
mice 114, 115
Michaelmas daisy 96
Miller, Jeff 146–7, **147**
monoculture 105–6
motorway embankments 14
moving bees and hives 48–51

N

'natural' beekeeping 33
Natural History Museum **14**, 15
nectar 19, 35, 53, 55, 65–6, 86–7, 102
    colour 91
    flow 89
    foraging for 71–2, 91–2
nectar substitute 86–7
New York 8, **10–11**
nosema 114
nucleus
    box **15**, 39, 40, **43**, 79
    buying 36, 38–9, 78
    hiving **15**, 40, 42, **43**

O

O'Callaghan, Martin 136
oil-seed rape 105
open mesh floor **29**, 31
orientation flights 72
ownership, legal status 45

P

packaged bees 44, **46**, **47**, 49–50, **149**
Paris 8, **9**, **19**
parks 14, 15
Paul, Johannes **6**
pear 88
pesticides 95
pests 31, 113–14
pets 17
phacelia 96
pheromones 19, 40, 41
photinia 78, **82**
pine hives 27
Pisano, Tony 122–4, **123**, **124**
plants for bees 85–97, 105–6
plum 88
pollen 19, 35, 53, 55, 66, 70, 85, 86, 91

beekeeping season 85–6
    colour 71–2, 86, 87, 88, 91
    foraging for 71–2, **73**, 91–2
pollen sacs 71, **73**
pollination corridors 92
pollinators, bees as 14, 16–17, 19, 26, 35, 53
poppy 96
predators 114–16
propolis 91
pupae 55, 66, 70

Q

queen 19, 31, 36, 53, 55, 61
    buying 49, 50
    checking 59, 61, 67
    finding 42, 59, 61
    hiving 42
    life cycle 75, 79
    marking 42, 45, **45**
    mating 81
    new 79, 81
    packaged bees 44, **46**, **47**, 50
    replacing 50
    splitting the hive 38, 78–9

swarming bees 38,
  74, 79
  virgin 79, 81
  winter months 113
queen cups 79
queen excluder **29**,
  32, 66, 72, **100**, 134
  removing 61
  smoking through
  60, 61

R

railway
  embankments 14
raspberry 91, 92, 96
refractometer 103
regurgitation 91
Roberts, Elvin 127–8,
  128
Rock, Ted and Valerie
  151–3
Root, Amos Ives 24–5
rose **89**
rosemary 88, 89, 106
royal jelly 79, 86

S

Scotland 25, 105
scout bees 39, 74, 89
security 12
sewage works 14
Shannon, Fiona
  156–8

Shiga, Yuichi 125–6
Siberian squill 87
Simon, Charles
  Martin 134–5
skeps **21**, 23
smoker 40, 57–9, **59**
  smoking bee stings
  40, 41
  smoking through
  queen excluder
  60, 61
snowdrop 86
South Africa 166–7
space requirements
  17
Spain 8
Speke Hall 154–5
spinner 101–2, **108**
splitting the hive 38,
  78–9
spring 53, 69–72, 85,
  96
stings 39–40, 41
  allergic reactions to
  41
sucrose 102
sugar fondant 113,
  117
sugar syrup 87
summer 55, 96
summer bees 70–1
sunflower **94**, 95, 96
supers **29**, 32, 72
  adding **76**

weight 99, 101
swarms 36, 74–9, **79**
  catching 36–8, **37**,
  81
  swarm control 77–9

T

tapping 125
thyme **87**
transporting bees
  48–51
Turner, Liz 128–9

U

uncapping 101–2, **108**
United States 44–8,
  105, 122–4,
  130–5, 142–3,
  146–53, 159–65

V

vandalism 12
Vane, Nikki **57**, 61
varroa mite 31, 114
veil 40, 56, **57**
venom 40, 41
viper's bugloss 96

W

Ware, Caroline 15
Warré, Abbé Émile 25

wasps 114, 115–16
water supply 14, 17,
  95
wax 23, 26, 72–3
  caps 66, 91, 101–2,
  103, **104**
wax sheets 73
White House beehive
  84–5
wild bees 19, 21, 23,
  32, 72–3, 126
Williamson, Jacque
  130–1
willow 85, 86, 96
wind, protecting
  against **30**
winter 35–6, 53, 70,
  81, 83, 113–19
  plants 96
  wax flakes 101, 116
winter bees 70–1, 81,
  83, 86, 94
winter cherry 96
wintersweet 96
woodpeckers 114–15
workers 19, 31, 36,
  53, 70
  life cycle 75, 81, 83

Acknowledgements

Many fellow beekeepers gave
generously of their time, help,
and advice as I wrote this book,
none more so than John Chapple,
Nikki Vane, and Barnaby Shaw
in London, and Keith Blignaut in
South Africa. Howard Nichols, a
master beekeeper, and Liz Turner,
a beekeeping novice, read through
my manuscript and offered sage
comments. Others patiently
provided many of the photographs
without which the book would
be infinitely less attractive. The
subjects of the case studies took
great trouble to answer all my
questions with care, consideration,
and enthusiasm. Anna Mumford
and Franni Farrell were my
wonderfully supportive editors at
Timber Press. To all, my thanks.

Photo acknowledgements

Chris Albert 132, 133
Kathryn Archer 57
Karl Arcuri 46, 47, 66 right, 148, 149, 150
Tim Baker 144, 145
Sándor Balázs 36
Mark Barton 31, 43 bottom right, 58 middle, 60, 77
Jason Bathe 100
Phil Beard 107
Eugenio Bono 174, 175
Kristian Buus 2, 12, 16, 50, 51
Canadian Opera Company 139
Patricia Choi 80
Greg Corman 160, 161
Mike Cornfield 10, 68, 87, 94
Michal Czyz 90
Tom Davies 155
Fairmont Waterfront 140, 141
Noel Faucett 14, 15 bottom, 27, 43 all except bottom right, 58 left and right, 59 left, 66 left, 67, 78, 104, 105, 120
Kristen Finn 147
Kelly Fitzsimmons 163
Felipe Fonseca 103, 108, 109, 110
Garfield Park Conservatory Alliance 152, 153
Herman Hooyschuur 21
HK Honey 170, 171

Vera Johnson 143
Katy Jones 98
Erik Knutzen 135
Amadis Lacheta 137
Orest Ladyzhynsky (hartfordphoto.com) 84
Omlet 20, 25
Marian Ovidiu Viorel 86
Janice Perry 15 bottom, 24 middle, 76, 82
Leila Pisano 123, 124
Nick Pulos 131
Jane Sebire 49, 71
Fiona Shannon 93, 88, 157
Yuichi Shiga 125
Mike Slocombe 165
Ludmila Smite 112
stock.xchng 94 left
Pascal Thauvin 89
Siobhánn Tighe 22, 121 right
Amanda Tobier 165
Eric Tourneret 6, 8, 9, 18, 35, 169
Liz Turner 25 middle and right, 37, 127, 128, 129
Richard Twilton 5, 52, 54, 73, 158
Nicoletta Valdisteno 59 right, 62, 64
Daan Verhoeven 13, 95 bottom, 97
Jacque Williamson 24 left, 118, 130
Viviene Yip 172